The Sheep Farming Guide

FOR SMALL AND
NOT-SO-SMALL FLOCKS

THE SHEEP FARMING GUIDE

for small and not-so-small flocks

Clive Dalton & Marjorie Orr

Foreword by John Gordon

First published 2004

ISBN 1-877270-72-5

Published by Hazard Press Limited
P.O. Box 2151, Christchurch, New Zealand
Email: info@hazard.co.nz
www.hazardonline.com

Illustrations by Mark Henderson

Printed in New Zealand by Astra Print

Contents

Foreword

Of all the resource books on farm livestock I've come across, *The Sheep Farming Guide* is the most down to earth and accessible. It aims to inform and assist shepherds with little or no experience of sheep farming. It succeeds admirably, both in terms of the quality and reliability of its information and its readability.

The original intent was to provide information for lifestyle farmers in particular but that was far too limiting. This book is now the perfect lifestyler's companion, but whether you have 5 sheep or 5000, the basis of sheep management remains the same. Every farm cadet or trainee should have a copy of this text to answer questions the boss may have glossed over or expected him or her to know already. What's more, I would defy any experienced sheep farmer not get something useful from this book.

I have worked – and in many ways wrestled – with sheep for most of my life: as an agricultural student, shepherd, farm manager, rural journalist and sheepdog triallist and commentator. Yet I learnt a tremendous amount from reading this book. It is essential reading also for anyone in the agricultural servicing industry who wants to understand their clients' language and business. And it would make a perfect wedding present for other inexperienced arrivals to farm life!

The book contains information on the most practical matters that, physically and mentally, all sheep farmers have to grapple with. Even catching, holding and lifting sheep are all dealt with in easy-to-follow passages. So too are other mysteries of ovine life: parasites, diseases and everyday preventative care. The book contains a fully itemised annual calendar of management, from dagging to dipping, tagging to tupping.

Naturally enough, this book succeeds because of its authors. For many years as a diagnostic veterinarian Dr Marjorie Orr has communicated to farmers what was wrong with their sheep and why; her knowledge continues in these pages. The first author, Dr Clive Dalton, I have known for many years and have always been in awe of his ability as a communicator. Clive is an agricultural scientist who not only knows what to say but how to say it. In my opinion he has no equal in this country.

As Clive and Marjorie state at the beginning: 'Sheep are likeable – you are allowed to like sheep!' Your enjoyment and management of your flock will be greatly enhanced by their book, which will also provide a wealth of information for many years to come.

John Gordon
(Rural wordsmith)

Introduction

Most of us are farmers at heart. Perhaps it's in our genes. After all, if we look far enough back we all find generations of farmer ancestors who lived largely on what they could grow on the land. Many of us still have the urge to get our hands dirty, to manage the land, to farm livestock and to provide at least some of the family's food, and we do it because of the satisfaction it provides.

This dream is becoming a reality for more and more people. A few years ago it was estimated that there were over 100,000 small farms in New Zealand, and the number is still growing. Why should this be? Maybe it's partly because transport to and from the city is relatively quick and easy. And certainly part of the reason is that modern technology has made it possible for many of us to work at home using our computers. More and more people are making their home close to the city rather than in it, so they can farm while continuing to enjoy the benefits of city life.

For many if not most of those who make that choice, farming sheep is part of that lifestyle.

Why sheep? Because they are small, gentle and placid, they have few and simple needs and there are lots of them so they are usually easy to get. They may be perverse and a little uncooperative from time to time, but if you know what you are doing and farm them well, you will enjoy sheep farming and it will be profitable. If you don't know what you're doing, it is easy to make a real mess of it.

Rest assured there will always be neighbours looking over the fence to complain when the farmer gets it wrong! A disproportionate number of animal welfare complaints are directed at lifestyle farmers – and many of them relate to sheep. They may be very dirty at the back end, or maybe they haven't been shorn for years, or are obviously sick. Often there is no deliberate cruelty on the part of the owner, nor any deliberate neglect. Lack of knowledge has resulted in a welfare problem that could easily have been prevented.

The fact that you are reading this book means you have taken a significant step toward becoming a better sheep farmer. Here you will find easy-to-digest information about setting up as a sheep farmer, managing the flock and preventing and controlling disease.

We have worked with sheep for years: Clive as a sheep researcher and teacher based in Hamilton, and Marjorie as a veterinarian and welfarist based in Dunedin. But we would like to acknowledge that much of our collected wisdom, such as it is, we owe to our best teachers – the experienced sheep farmers who have so willingly shared their expertise with us. Unfortunately they don't write books, so we are privileged to be able to pass on much of their teachings to you. We want to share this knowledge so that you can enjoy farming sheep too.

Our biggest vote of thanks goes to a lifestyle farmer who, a few years ago, told us he wanted to buy a few sheep and then innocently asked what he needed to know! We started to make notes … then more and more … and this book is the end result.

So read on – learn how to enjoy being a 'sophisticated peasant sheep farmer' and make a profit at the same time.

Clive Dalton and Marjorie Orr

PART ONE
Getting Started

1. Why Keep Sheep?

Points in favour

- They are likeable – you are allowed to like sheep!
- Sheep are small placid animals, and relatively easy to farm.
- They suit almost any property – even steep paddocks and poorer pastures.
- They can keep small areas such as orchards or driveways grazed down.
- They don't produce large quantities of faeces the way cattle do.
- They don't pug paddocks in wet weather the way cattle do.
- Pastures grazed by sheep have a denser sward with fewer bare areas.
- The commercial market for sheep products (meat, wool and milk) is usually fair to good.
- Your family can be self-sufficient in meat and wool.
- Sheep require less capital to buy and maintain than cattle.

Points against

- They need good husbandry for optimum production.
- You need good fences.
- You need good yards to handle them easily.
- Most sheep should be shorn once a year.
- When pasture is lush you have to dag and crutch them

regularly to keep their rear ends clean.

- Dog worrying can be a problem, especially near urban development.
- You will probably have to treat them regularly for internal and external parasites.
- In many parts of New Zealand you have to provide cobalt/vitamin B12 and/or selenium supplements.
- Feet may need to be trimmed from time to time to remove excess horn and avoid footrot.
- Flystrike is a risk throughout summer.

So what are sheep?

Sheep are ruminants, which means they have three fore-stomachs as well as a true stomach, and they chew their cud like cattle. Like all ruminants they have no upper incisor teeth but sheep can graze short pasture more effectively than cattle because they can bite closer to the ground.

Sheep are social animals that tend to 'flock' together for security. A sheep on its own (unless it's a pet lamb) will be very stressed – so always provide company. They can remember individuals in the flock for years, even after separation. They sort out a social order by head butting.

Their breeding season begins with mating in autumn, triggered by shortening daylight length. When rams become sexually active they develop a strong odour – but not quite as strong as billy goats! Rams will form harems of ewes if they are allowed to.

Multiple births (twins and triplets) are common in most breeds of sheep if they are fed well, but breeds vary in their inherent fertility. A strong bond develops between ewe and lambs with frequent spells of suckling.

Sheep are a 'follower' species – i.e. lambs tend to stay near ewes and follow them. This contrasts with cattle, deer and goats, whose young 'lie out' away from their dams for long periods each day, with short intensive bouts of suckling.

Sheep store fat both internally and under their skin. Their wool acts as insulation in both cold and hot conditions. They have acute hearing, a good sense of smell and narrow binocular vision directly in front of them. They also have wide peripheral vision to the side and rear but cannot see in the area immediately behind them. Although it appears sheep can distinguish certain colours, this has not been completely researched.

Sheep seek out high ground for security – they like going uphill.

They see the dog as a major predator unless they have been reared with dogs.

The history of sheep in New Zealand

Captain James Cook landed the first two sheep (Merinos) in Marlborough in 1773, but they both died before producing offspring. In 1834 the first major shipment of sheep (Merinos from Australia) landed on Mana Island. These sheep were later moved to the Wairarapa, and from 1840 more Merinos were imported from Australia as more land was taken up for grazing.

Between 1850 and 1860 several English long-wool breeds (e.g. Lincoln and Leicester) were imported with the early European settlers, and crossed with the local Merinos. These British breeds proved better suited to high-rainfall areas than the Merinos, and the crossbreds produced what were called 'colonial halfbreds', with improved fertility, wool and carcases.

A turning point in the industry occurred in 1882 when the first shipment of refrigerated mutton left New Zealand for Britain. After this, farmers changed to breeds that produced more meat than Merinos, which then declined in number and retreated to the drier high country.

The strong demand for sheep meat and wool after the Second World War caused a great expansion of the ewe flock, and also the development of new, more productive breeds for hill country such as the Coopworth and Perendale.

Sheep production was further encouraged by government support until the mid-1980s, when all subsidies were withdrawn. New Zealand's sheep population peaked at this time at around 70 million, but since then sheep numbers have declined to around 39 million in 2002 as a result of declining markets for both meat and wool.

The improved demand for sheep meat (not for wool) in 2000, together with the low value of the New Zealand dollar, improved sheep profitability considerably after many years of depression.

Prospects for sheep meat at the time of writing are very positive.

What do sheep give us?

It's well known that sheep produce meat and wool, and now there are a few commercial farmers milking sheep for cheese and other milk products.

But many other usable products are produced from sheep, including blood, bones, stomach and intestines, heart, spleen, tongue, glands and fat.

The natural grease from wool (lanolin) is also widely used.

Are they really stupid?

Sheep are not the dumb animals many people take them for. They are gentle and harmless and they like being with other sheep – and that's not so stupid.

Sheep are capable of learning simple routines like coming when called, finding holes in fences and opening gates, and they can learn these tricks from one another. Food rewards are the way to teach sheep routines and tricks, if you think it's a wise move. You may live to regret it!

Lambs quickly learn from their dams such things as eating new feed (concentrate meal, grain, hay, silage), cracking open chestnuts with their feet, selecting garden flowers and so on.

Stupid? We don't think so.

2. What's in a Name?

What is a two-tooth? A wether? A cryptorchid? There's a lot of sheep jargon that can easily snag the uninitiated. Before we go any further, here's a beginner's guide to some of the terminology.

It's all in the teeth ...

You can estimate the age of sheep by their lower incisors, because the temporary incisors are gradually replaced by permanent incisors in pairs, working from the middle outwards. (They have no top incisors.) Sheep are often named for the number of teeth they have. Bearing in mind that there is enormous individual variation between sheep, here is what you normally see:

- **Lambs** (up to nine to 10 months old) have eight temporary incisors (milk teeth).
- **Hoggets** are older lambs in which the central pair of permanent teeth have not yet appeared. Lambs become hoggets in their first winter.
- **Two-tooths**: The central permanent pair of teeth start to appear at about 12 months of age and are fully erupted before 18 months of age.
- **Four-tooths**: The second pair appear at 21-24 months old.
- **Six-tooths**: The third pair erupt at 30-36 months of age.
- **Full mouth**: The sheep has a 'full mouth' when the set of

eight permanent teeth is complete – 42 to 48 months of age. After four years of age it can be difficult to assess age because of tooth wear or lost teeth.

Other sheep names

Alpha lamb: An unweaned lamb too heavy for the beta trade, up to 18-20kg liveweight. Not usually tail-docked or castrated.

Bobby lamb or **beta lamb**: A lamb not yet weaned (one to three weeks old) going for slaughter at around 14kg liveweight. Not usually tail-docked or castrated.

Cast sheep: A sheep found lying on its back unable to get up, usually because it is heavy in lamb or has a heavy fleece.

Cast-for-age (CFA) ewe: An old ewe (usually over five years) culled from the flock because of age.

Cryptorchid: A true cryptorchid is a male with undescended testicle or testicles, but the term is also used to describe a lamb that has been made infertile (i.e. castrated) by applying a rubber ring below the testicles to remove the scrotum, leaving the testes pushed up against the body (short scrotum technique).

Cull sheep: One selected for slaughter or sale.

Dry ewe: One that did not produce a lamb.

Dry/dry: One that did not produce a lamb because it did not get pregnant.

Ewe: A mature female sheep, usually over two years old.

Flock ram: A non-registered ram used in a commercial flock.

Long-tailer: A male whose tail was left undocked to indicate that it wasn't castrated.

Prime lamb: A lamb that is ready (finished) for sale to a meat company or fat-stock buyer. It is a more acceptable term than 'fat lamb'.

Ram: A male sheep of any age.

Slink: A lamb that was either born dead or died soon after birth. Slinks are processed for their skin.

Stud ram: A purebred ram registered with a breed association.

Tail-up or **chaser** or **follow-up ram**: A ram used at the end of mating (joining) to mate any late-cycling ewes.

Teaser: A vasectomised ram put in a paddock near ewes to encourage them to cycle prior to mating.

Terminal sire: A ram used to sire lambs for the meatworks (as opposed to replacement stock). It's the last ram used in a crossbreeding programme, so all the lambs produced go to slaughter.

Wet/dry: A ewe that produced a lamb but didn't rear it (perhaps it died).

Wether: A castrated male sheep.

3. Choosing a Breed

There is a wide choice of pure breeds available in New Zealand. Some of these have been developed by merging other breeds. Your choice may be based on availability, what you want to produce from your flock (meat, wool or milk), or personal fancy. Here are the breeds of sheep currently farmed in New Zealand.

Meat breeds

Awassi (fat-tailed)
Dorper
Dorset Down
Dorset Horn
Hampshire
Oxford Down
Polled Dorset
Ryeland
Shropshire
South Dorset
South Dorset Down
South Hampshire
South Suffolk
Southdown
Suffolk
Texel
Wiltshire

Wool breeds
Boroola Merino (fine wool)
Drysdale (coarse wool)
Merino (fine wool)
Polwarth (fine wool)
Tukidale (coarse wool)

Dual-purpose (meat and wool) breeds
Border Leicester
Borderdale
Cheviot
Coopworth
Corriedale
Dohne
East Friesian
English Leicester
Finnish Landrace
German White-faced Mutton
Lincoln
Perendale
Romney

Milking breeds
East Friesian
Polled Dorset

Pelt breeds
Gotland Pelt
Karakul

Rare breeds
Arapawa
Hokanui Merino
Pitt Island Merino
Raglan Romney

Table 1. Production characteristics of various breeds

Breed	Body weight (kg)	Fleece weight (kg)	Sample (fleece) length (mm)	Fibre diametre (microns)	Lambing %
Merino	35–45	3.5–5	65–100	19–24	70–100
Lincoln	55–65	5–7	175–250	39–41	100–120
Romney	45–55	4.5–6	125–175	33–37	85–120
English Leicester	55–65	5–6	150–200	37–40	100–120
Border Leicester	50–60	4.5–6	150–200	37–40	110–130
Cheviot	40–50	2–3	75–100	28–33	90–110
Corriedale	45–55	4.5–6	75–125	28–33	90–120
NZ Halfbred	40–50	4–5	75–110	25–31	85–100
Perendale	40–50	3.5–5	100–150	30–35	90–120
Coopworth	50–60	4.5–6	125–175	35–39	110–130
Drysdale	45–55	5–7	200–300	40+	90–120
Poll Dorset	50–60	2–3	75–100	27–32	110–130
Dorset Horn	50–60	2–3	75–100	27–32	110–130
Southdown	50–55	2–2.5	50–75	23–28	100–120
Hampshire	55–65	2–3	50–75	26–30	100–120
Dorset Down	50–55	2–3	50–75	26–29	100–120
Suffolk	50–60	2.5–3	75–100	30–35	100–120
South Suffolk	50–60	2–3	50–75	28–32	100–120
Borderdale	50–60	4.5–6	100–150	30–35	110–130

Purebreds

When rams and ewes of the same breed are mated they produce purebred offspring.

The ewe offspring are kept for breeding and rams of the same breed are bought in to mate them. Ram offspring may be sold for meat or as stud sires to other breeders.

Farms that sell purebred sheep (usually rams) are called stud farms. These farms are registered with the appropriate breed association or society and their sheep are registered in a stud flock book.

Crossbreds

When rams are mated to ewes of a different breed, the offspring are called crossbreds.

Crossbreds are generally more productive than purebreds because they usually have 'hybrid vigour' or 'heterosis'. This hybrid vigour can be very variable but on average the first-cross offspring are significantly more productive than the *average* of the parents. This means that generally their survival rate as lambs is better, they grow faster, they are more resistant to disease and they produce more lambs.

Crossbreds also often fare better than purebreds in harsh environments, e.g. on dry hill country.

The extra production with hybrid vigour can be as much as 10-13% in first-cross offspring, but it generally declines with further crossbreeding.

Maintaining hybrid vigour in your flock

By buying in top sires of outside breeds to mate with your crossbred ewes, you can often keep a fair amount of hybrid vigour going.

The wider the genetic differences between the breeds crossed, the greater the hybrid vigour. If you cross two breeds that are very similar you can't expect much hybrid vigour. You will end up with offspring that are an average of the parent breeds.

The simplest way to maximise hybrid vigour in your flock is to keep using the purebred parents, and send all the first-cross lambs to the meatworks. When you want to replace the purebreds, buy them in again.

Alternatively you could produce replacement ewes and rams yourself, but you'll need flocks of each breed – a larger flock to breed the ewes and a smaller ram breeding flock. Keep the first-generation (F1) females as dams for further breeding. These F1 females are generally fertile and good milkers, so their lambs grow well to weaning. Mate them to a terminal sire (a ram used to sire lambs for the meatworks), and send all their progeny for slaughter – they'll have excellent heavy carcasees.

Interbreeding

If all this sounds too complicated, then you can forget about what stage of crossbreeding you are at, and just keep the best ewe and ram lambs to become your dams and sire/s.

This is called 'interbreeding', or breeding within the population or 'breed' you have, although a purist would consider you shouldn't use the word 'breed' for the resulting population of 'crossbred mongrels'!

But that's how New Zealand breeders made the Corriedale (Merino x Romney), Perendale (Cheviot x Romney), and Coopworth (Border Leicester x Romney). They started talking about their crossbreds' 'breed' after about the third cross (F3), and became stud breeders. They were then respectable again!

Inbreeding

Inbreeding occurs when you mate sheep that are related or have ancestral genes in common. This increases the chances of identical genes pairing up and reduces genetic variation, which is valuable as a basis for selection and improvement.

Genes that are common to both parents may be good genes, so inbreeding can be good in concentrating good traits in the progeny. But they may also be bad genes, causing congenital diseases and defects such as undershot jaw or short legs. Other inbreeding defects include reduced fertility, deformities, lowered lamb survival and lowered resistance to disease.

If inbreeding is intense, e.g. a sire is mated to his daughters or a son back to his mother, then the defects seen may be more serious and may appear with greater frequency. Make it a rule to change your rams before there is a risk of daughters being mated by them.

If you must mate relatives, make sure they are as distant as possible. Don't mate any animals closer than first cousins. At this level you could call it 'line-breeding' i.e. mating relatives, but at a low intensity.

To delay the build-up of inbreeding in a small flock, form the ewes into groups (families) and use rams from each family to mate another family in a planned rotation. This will extend the time you can use each ram.

4. Starting a Breeding Flock

What to buy

It can be difficult for the inexperienced farmer to know what to buy – there are so many options. Seeking help from an experienced sheep farmer is a good start, because what you should buy depends on the type of farm you have as well as your personal likes.

It's generally best to run a 'mixed-age' ewe flock (with sheep of all ages), and it's a good idea to buy them at these different ages to have an 'age-balanced' flock.

You really want four age groups of productive ewes in a flock. These are two-tooths, four-tooths, six-tooths and full-mouth ewes. The age balance is maintained by replacing the five-year-olds each year with two-tooths going into the breeding flock.

Older ewes are always more productive, with maximum productivity at about the four-tooth stage. At about five years of age their productivity falls, when they start to lose their teeth or the teeth get worn. They may also have udder problems such as pendulous udders, dry quarters from mastitis infections or blind teats that don't work.

But if you manage your sheep well on a small block, there's no reason why they shouldn't live well beyond five years old. Plenty of sheep reared for school pet days have lived for 12 years or longer and produced lambs each year. We even know of a few who have made it to 19 years – but they have been well loved and very well fed!

Hill-country cast-for-age (CFA) ewes are sometimes sold to lowland farms for crossing with a meat sire. These can do very well on lifestyle farms when they are well looked after.

If for some reason you have to cull the older age groups heavily, then mating hoggets can help boost replacement numbers and save you buying in as many sheep. But remember, hoggets need very good care, as you are asking young growing animals to work like mature sheep.

Where to buy

Get your ewes from a reputable breeder who has surplus stock for sale. They may be 'cull' sheep, but a good breeder will guarantee the animals for you, and replace them if they prove defective. Beware of cull sheep at sales if you know nothing about them. The auctioneer may make them sound a lot better than they are.

At various times of year specialist sheep fairs sell lambs, two-tooths and five-year-old ewes – all guaranteed by the vendor or stock firm as being 'sound in mouth and udder'.

Sometimes there's a flock dispersal sale where all age groups are up for auction. These are often in large numbers but you may be allowed to select a few from each pen to get the number you want. Buy a few extra two-tooths to allow for wastage along the way – if some of the younger sheep are barren, for example.

Which ewes to replace

Autumn is the best time to make decisions about the flock's future, before the rams go out. It's tidy-up time before winter, as you don't want to waste feed by keeping unproductive animals.

In a mixed-age flock, normally all the five-year-olds would be marked for culling, but you may wish to keep the best of them if other age groups are down on numbers.

Unless they lambed as hoggets, two-tooths will have had their first lambs and they will have had the hardest time. Some will have been wet/dry (i.e. their lamb died) and a few will have been dry/dry (they did not have a lamb). Be wary of keeping any of these sheep in case they perform like this again. Their problems may be genetic and not just bad luck.

Any other ewe you keep (apart from the two-tooths) should have successfully reared a lamb or lambs that season. Be ruthless about this: your flock's fecundity (ability to rear lambs) is the key to your profit. You may be a bit more lenient about a hogget that did not lamb successfully.

The following are good ewes to keep:

- Ewes (especially young ones) that have reared their lamb or lambs to weaning without assistance, and whose lambs had above-average weaning weights.
- Ewes that have reared twins unassisted. Birthing problems can be partly hereditary.
- Ewe lambs that have been born and reared as multiples (twins or triplets).
- Ewes that have had multiple lambs and reared them every year. This is building fertility and fecundity into the flock.

If you are interested in wool, record hogget fleece weight and keep the heaviest wool producers with wool that is true-to-type for the breed. Don't bother to record later fleece weights – hogget fleece weight is a good indicator of later fleece production. Cull any sheep with fleeces that are not true to breed type. Wool quality can be hard to assess so get some good advice from an expert.

Which rams to use

Buy your rams from a breeder who:

- Has a flock you like and has similar objectives to your own.
- Supplies comprehensive performance records and is willing to explain them.
- Will help you select a ram.
- Will explain their farm's breeding and selection programmes.
- Will listen and understand your flock needs.
- Will guarantee any rams you buy.
- Is making genetic gains. If the breeder is in an official recording scheme, he or she will have records to demonstrate the gains being made.

Physical characteristics

Before you buy in any sheep or decide which of your own young ewes to select for your breeding flock, check for obvious faults to make sure you keep only healthy sheep that are likely to do well.

Teeth. In the lower front jaw there should be eight good firm incisors.

Jaws. The lower incisors should meet the hard top dental pad just behind its front edge – not in front of or behind it.

Udder. A ewe's udder should be healthy, with no hard lumps or damaged teats, and not pendulous (hanging down as a result of stretched ligaments). There should be two well-placed teats that are not damaged.

Feet and legs. The sheep should have normal-shaped feet and legs with no footrot, abscesses or lameness.

Wool. The wool should be typical of the breed, with even cover over the body. Avoid sheep with excess wool around the eyes or down the legs.

Colour. This should be typical of the breed. Black fibres in white breeds are considered a serious defect in the wool trade.

Dags. It can be wise to cull sheep with a lot of dags as this trait tends to be inherited and predisposes the sheep to flystrike. It may indicate low resistance to internal parasites.

Running a non-breeding flock

Who says your sheep need to have lambs? If you don't want the work of organising mating and lambing, you could consider these options:

- Buy in wether lambs or cryptorchids (lambs castrated by short-scrotum technique) at weaning (four months old). Grow them well and get them to the meatworks in winter before their first two permanent teeth erupt, so they'll still be classed as lambs for the meat trade.
- Buy in ram lambs at weaning, feed them well and sell them as lambs for the meat trade before their first pair of permanent teeth erupt.
- Run a flock of older wether sheep. These can graze surplus pasture, produce wool each year, and mutton when finally

culled. In good times they'll put on fat, and in hard times you don't need to worry if they get a little bit leaner. Usually they are easy to manage and come running to the sound of food pellets rattling in a bucket!

- Run a flock of ewes (any age) and don't mate them. Treat them like a wether flock for grazing control. But make sure a ram doesn't sneak in among them in autumn – a lot can happen in one night!
- Buy ewe lambs at weaning and feed them so that they grow well, then sell them about a year later, ready for mating to lamb as two-tooths. They need to be well cared for to make good money. If you feed them extra well they may mature enough to be sold after a few months ready for mating (so they lamb when they are one year old).

5. Alternative Husbandry Methods

'Easy care'

'Easy care' refers to a system of sheep management that involves minimal intervention by the farmer. It's most relevant on larger hill farms. Properly carried out, it means that the sheep are farmed in such a way that they don't require much assistance.

Sheep that have been bred for self-sufficiency, on the right type of hill land with a low stocking rate, are sturdy, they get plenty of exercise over varied terrain, sufficient feed and shelter, they generally don't require assistance to lamb, they are very good mothers, they have only one or two lambs at a time, they don't have bearing problems and they have strong feet.

They must be farmed in conditions that do not allow a big worm burden to build up on pasture, and which do not expose sheep to the wet, muddy conditions that might cause foot problems. At lambing they should have access to an area relatively free of hazards such as steep hills and ditches, and they must have access to shelter in bad weather.

'Easy care' does not mean neglecting sheep. The farmer has to keep a watchful eye on his or her flock and may inspect them daily, drenching as appropriate and crutching and shearing once a year.

'Easy care' sheep are bred by culling from the flock any ewe that is disease-prone or that does not successfully raise a lamb to weaning for whatever reason.

Sheep that have been bred to be self-sufficient on hill blocks may not be suitable for improved blocks. This means that the 'easy care' style of farming is generally not appropriate for small holdings or small farms. The better feed on offer before mating on small farms can lead to more twins and triplets, with more risk of bearings in late pregnancy, mis-mothering and smaller, less robust lambs, and it can also mean bigger single lambs with an increased risk of lambing problems.

On the softer, improved land of many smaller farms, feet may not wear as they do on the hill so they may require trimming, and wetter muddier conditions can predispose to foot problems.

Going organic

There has been a massive interest in organic farming in recent years, driven by the demand for so-called healthy food by overseas and domestic consumers. However, farming sheep organically demands greater management skills than farming conventionally.

Some sheep farmers are attracted to organic farming because they want to reduce the amount of chemicals used on their farms and also the cost, or because they want to produce what they believe is healthier food for their family, or because they want to cause less damage to the environment.

Do your homework before you start

Organic farming is a major undertaking so you must do your homework before you convert.

Visit organic farms. This is the best way to get knowledge as there is little written in books about farming sheep organically.

Visit farmers at different stages of conversion to organics so that you are aware of the problems that can appear at the start, and you know how to prevent them and deal with them if they occur.

Talk to a field officer from one of the various organic certification organisations. They are very practical people who have helped many farmers 'go organic' and they know all the problems ('challenges'!) involved.

They will sell you their organisation's manual, containing detailed information on how to go about it.

Conversion generally takes from two to three years.

Organic feed

Healthy stock eat healthy pastures grown in healthy soil – this is the basic thesis of organics. If you buy in supplementary feed for your sheep, it must also have been grown organically, or a high proportion of it should have been.

Organic supplementary feeds can be difficult to find and cost more.

Animal health

This is the most challenging area. Great emphasis is placed by organic farmers on preventing disease by having management systems that feed stock well and reduce stress.

Breeding sheep for resistance or buying sheep that have a relatively high resistance to diseases such as footrot and gastrointestinal parasitism has high priority.

Homoeopathic remedies are commonly used for treatment of illness or disease where it does occur.

However, if such treatment does not produce results, sheep should be treated with conventional remedies to prevent distress and suffering.

Under organics, after treatment with a conventional animal remedy, e.g. an antibiotic, sheep must be held in a quarantine area for twice the meat withholding period for that remedy before they are returned to the organic area. These sheep then lose their organic certification for a designated period, generally 12 months.

Specific diseases

- Use of vaccines against the clostridial diseases ('5 in 1' vaccine) and scabby mouth is allowed if you can show there is a need for the vaccination on your farm.
- Footrot: Zinc sulphate and copper sulphate footbaths are allowed for prevention and treatment of footrot and foot scald.
- Internal parasites: The focus is on prevention through good pasture management (see Chapter 20). There are commercially available organic drenches.
- External parasites: Vegetable oils and sulphur preparations can be used for treatment.

- Flystrike: Neem and tee tree oil, garlic and pyrethrum can be used to treat lesions.
- Facial eczema: Zinc oxide ointment can be used to help prevent and treat skin damage. As a preventative, administration of zinc oxide by oral drench or rumen bolus may be allowed.

Contacts

BIO-GRO New Zealand, PO Box 9693, Wellington.
Website www.biogro.co.nz.

Biodynamic Farming and Gardening Association in NZ
PO Box 39045, Wellington.
Email biodynamics@clear.net.nz

AgriQuality New Zealand Ltd
PO Box 98905, SAMC, Auckland.
Website www.agriquality.co.nz

Pet sheep

These are usually well-loved sheep, but sometimes they are over-indulged and become obese. This makes them prone to foot problems, arthritis and getting cast on their backs.

If you use them as mowing machines to graze rough areas, don't forget to give them all the attention other sheep get – water, foot trimming, drenching etc.

If they have access to the garden (and let's face it, many pet sheep even have access to the living room!), make sure they don't eat poisonous plants such as rhododendron.

Keep their wool short so they don't get caught up in blackberry, or suffer heat exhaustion and blowfly attack.

Tethered sheep

If a pet sheep is tethered (and only a very tame pet should be tethered), make sure it is checked regularly, at least once daily, and remember that adequate feed and water must be available at all times.

Check that the collar doesn't chafe the neck. Leather collars are best, and there should be at least one swivel link in the chain. Don't forget to remove it before shearing or crutching.

Don't tether a sheep on a slope or near a ditch or anywhere its rope or chain could get caught up; otherwise it may be strangled.

Tethered sheep have little chance in dog attacks, they can be hit by cars and stolen for Christmas dinner – so think carefully about where to tether your sheep.

Remember too that sheep are social flocking animals. If a tethered sheep does not adapt to being on its own, then release it.

6. Fencing and Handling

Fencing

Sheep are best contained by a standard seven-wire, fully battened boundary fence, as laid down in the Fencing Act 1978. This type of fence will last for at least 20 years without need of much maintenance. But it is very expensive for internal fencing, where electric fencing can be used. If you have an old boundary fence and are short of cash, run an electric wire along it on outriggers to extend its life. Use trial and error to find the right height for the outriggers.

You'll need at least three hot wires at different heights to keep back both ewes and large lambs. Use insulated fencing stakes with many attachment points for the wire or tape, so you can adjust the wires to find the best heights for your sheep.

For mature ewes two hot wires will be effective. For ewes with young lambs it is better to use electric netting.

Sheep should be trained to respect an electric fence by putting a hot wire or tape around a paddock inside a conventional fence. Let them graze it out so they have to eat under the electric fence and learn.

Wool is a good insulator, so don't expect a hot wire to contain long-woolled mature sheep. Shear them regularly to keep the wool shorter than 100mm. Bare-headed breeds are easier to restrain than woolly-headed breeds. There will always be some sheep that have no respect for an electric fence. Cull them, as they'll lead and teach the others to escape.

Get rid of all barbed wire.

Tips for catching and holding sheep

Don't catch or hang on to sheep by their wool. It hurts. To catch a sheep, move up quietly in the blind zone immediately behind it.

To sit it on its rear for procedures such as feet trimming, put one hand under its chin and lift its head slightly to stop it lurching forward. If it gets its head down you will lose control. Place one hand under its chin, and turn its head around to face its rear on the side away from you. Grab its rear end with your other hand, or down where the back leg joins the body.

Hang on tight and move backwards, pulling the sheep toward you. The sheep's legs will buckle and it will fall back toward you. Lay it on the ground on its side and press down on it to keep it there. Then quickly grab its front legs and sit it up on its rear end at an angle of about 60 degrees from upright.

If it sits too far forward it will jump back on to its feet – be prepared for this. If it sits too far back it will struggle and kick with both back legs in unison.

Practise finding the right angle to sit the sheep at, and keep your legs close in behind its body. When you find the right angle the sheep will relax and you can take your hands off and hold it only with your legs. This is the position a shearer uses before starting to shear.

With a big heavy ram or ewe, you'll struggle to turn its head around as its neck will be too strong. Stand the sheep against the rails and kneel down beside it with your head in its ribs. Reach through below its belly and grab the two opposite side legs. Give the legs a firm pull toward you and the sheep should drop onto its side. Use that moment of surprise to hold it down, then grab its front feet quickly and sit it on its rump. Watch for flying legs when it realises what has happened!

Don't try to lift heavy sheep. When lifting a small sheep over a fence, rock it up and over on your knees to save your back. Pull the sheep up onto its hind legs, hold its right front leg with your left hand, grip the wool in its flank with your right hand, and use your right knee to push the sheep up and over the fence.

Take care not to injure your back. Try to keep your back straight at all times.

Tips for yarding sheep

To keep sheep moving, make sure there's a clear way ahead. Sheep following each other should see sheep moving ahead, preferably around a bend. Moving sheep will generally 'pull' the followers with them – once you've got a flow going.

They don't like visual dead ends and won't move freely toward them. Try to arrange things so they think they're about to escape back to their territorial area – the paddock they came from! If you have a dead end anywhere (e.g. in a race), put a mirror on the wall so they see a sheep to move to for security. Alternatively, pen a decoy sheep at the far end to help the flow.

Make races narrow enough to prevent sheep turning around and blocking the flow. This is not easy, as you have to handle sheep ranging in size from large pregnant ewes to small lambs in the same facilities. Having tapered sides to the race (slightly wider at the top than at the bottom) can help. Advancing sheep should not be able to see those following them, as they'll stop, then reverse or try to turn around.

Make sure the sides of pens and races where you do most of the handling are closely boarded so the sheep cannot see through and get distracted.

Sheep move best from dark into light, and generally dislike changes in light contrast. They don't like bright lights such as reflections from windows; neither do they like light coming up from under gratings. Gratings at woolshed doors should be laid so the floor looks solid to the sheep. Sheep really panic on slippery floors – so provide some grip.

They soon get used to any noise used to move them – so keep changing the noise for effect. Changing it (or stopping it) will also help prevent the helpers going silly!

Sheep remember past experiences. Run them through new facilities a few times and let them think they can escape before you subject them to any unpleasant procedures such as ear tagging or shearing.

If you have badly designed handling facilities that cannot be fixed, keep a 'Judas' sheep to lead the doubters through. Pet lambs are useful for this job. Train the Judas sheep with some pellets and cover it with raddle (marker crayon) to make sure it isn't accidentally loaded into the truck!

7. Legal Responsibilities of Livestock Owners

Animal welfare

Your legal responsibilities regarding animal welfare are set out in the Animal Welfare Act 1999. In summary, you must provide all your animals' needs: physical, health and behavioural.

It is an offence not to meet these needs 'according to good practice and scientific knowledge':

- You mustn't ill-treat any animal so that it experiences unnecessary or unreasonable pain or suffering.
- If you have a sick or injured animal you must treat it or get it treated. It is an offence not to alleviate its pain or distress.
- You mustn't kill any animal – even one you own – in such a way that you cause it unreasonable or unnecessary pain or distress.

The comprehensive *Code of Recommendations and Minimum Standards for the Welfare of Sheep* can be purchased from MAF, Box 2526, Wellington, for $5.

The code spells out animals' needs in terms of the 'five freedoms':

- Freedom from hunger and thirst
- Freedom from discomfort
- Freedom from injury and disease
- Freedom from fear and distress
- Freedom to express normal behaviour.

Animal control

You must not let your livestock wander onto public roads. Wandering stock are a real hazard and a common cause of accidents and you risk being liable for any damage caused. They are liable to be impounded by the local authority's animal control officers. You must maintain sound boundary fences to keep stock in.

If you have a pet dog or a working dog, you must register it with the local authority and its registration disc should be attached to its collar at all times. You must keep it under control at all times. This means it must stay on your property unless it is under the direct supervision of you or some delegated person.

If you find a dog wandering unattended or livestock wandering on the road, you should contact a local animal control officer.

If a dog wanders onto your property and you think it might be disturbing your sheep, you would not be penalised for shooting it, but you would have to justify your actions. Unless it is obviously worrying or threatening to worry your sheep, it might be more politic to take steps to remove it or have it removed by contacting the owner right away if you can.

If the neighbour's sheep wander onto your property, again it is best to try to resolve the problem amicably, but if this isn't possible you should contact your local animal control officers for advice.

If you witness ill-treatment of animals

Sometimes you may witness people neglecting or ill-treating their animals. It may be ignorance on the part of the owner, i.e. he or she may not know how to care for them properly. Or perhaps the owner is unwell or has major money problems and is not able to care for them properly. In these cases, unless the animal needs urgent attention, you might be able (tactfully!) to advise the owner what to do, and you might even be able to offer help if necessary.

If you don't know the owner, or he or she won't be helped, or if the animal needs urgent attention, you should act.

Deliberate cruelty to animals isn't common, fortunately, but it still happens.

Where the owner is deliberately causing his or her animal to suffer or letting it suffer, you must take action. The animal welfare helpline for MAF is 0800 327 027. Your local SPCA phone number is in the phonebook. These organisations have animal welfare inspectors who have the authority to investigate the problem and resolve it, without revealing your identity.

PART TWO
Basic Care

8. The New Zealand Sheep Farming Calendar

Every sheep farm is different and there are enormous differences between districts and regions in New Zealand. Bear in mind that North Island farms are at least a month ahead of South Island farms.

These are suggestions more than directions. Use this calendar as a 'memory jogger' and seek expert advice when you need it.

June

Livestock

- All non-productive sheep should be off the farm so that feed is allocated efficiently to the animals that need it.
- Check liveweight and condition score targets.
- Check that rams are being well cared for.
- Start planning for lambing – get your equipment and lambing paddocks ready.
- Check that all records are ready for lambing.
- Ensure that you have a disposal method for dead stock.

Feeding/Pasture management

- Carry out a feed budget to make sure you have enough pasture and supplementary feed.
- Check the feed allocations to ewes.
- Build up pasture feed reserves for lambing.

- After grazing, the residual pasture should be around 2-3cm high.
- In wet weather avoid pugging the pastures by removing stock or reducing stock density.
- Check the management of new grass paddocks. Plan for some nitrogen fertiliser at 25-30kg/ha when soil temperatures are above 6-10°C, provided soils are not too wet.

Animal health

- Discuss a detailed animal health programme for next season with your vet.
- Do blood profiles for minerals and trace elements if the vet recommends this.
- Carry out the pre-lambing vaccinations as recommended by your vet.
- Pay special attention to foot problems, especially on wet pastures.
- Watch out for early abortions and get all cases checked by the vet.
- Update your animal health records.

General

- A good time for general farm maintenance: fencing, drainage, swamps, creeks, stock crossings. You want this all up to scratch before lambing.
- Pay special attention to farm vehicles – especially bikes.
- Depending on your balance date, book the date for your annual formal review of the business with accountant, banker and farm consultant.
- Pay accounts monthly.
- Contact your banker/accountant to check GST payments.
- Estimate your cash position and bank requirements for the payment of creditors.
- Check Occupational Health and Safety (OSH) and farm safety policy for the farm.
- Attend farmers' conferences and vet practice training days.

July

Livestock

- Divide ewes into groups by lambing date to control feeding.
- Watch for sleepy sickness, especially in ewes that look to be carrying multiples and close to lambing.
- Separate early lambers and watch for any problems such as abortions or mastitis (red swollen udders and ewes limping on a back leg).
- Be prepared for ewes that lamb before expected date. Check them regularly.
- Have good feed and shelter ready for all ewes after lambing
- Put late ewes on short feed and keep an eye out for signs of lambing.
- Separate out ewes that are clearly barren (dry/dry) and either feed them less or dispose of them.

Feeding/Pasture management

- Keep a regular check on pasture growth and consider supplement use – feed sheep when you need to, to maintain stock condition.
- Consider the strategic use of nitrogen at 30-50kg/ha to provide feed for lactating ewes.

Animal health

- Update animal health records.
- Check the growth and health of any young stock (e.g. hoggets).
- Don't drench hoggets without first having a faecal egg count done.
- Vaccinate ewes before lambing – check with your vet.

General

- Check financial budget and cashflow.
- Finalise cashbook.
- Pay accounts monthly.

August

Livestock

- Lambing should be in full swing (North Island).
- Fully feed lactating ewes, especially those with multiples.
- Identify ewes into groups by lambing date to control feeding.
- Concentrate on saving every lamb. Watch out for the boredom that creeps in after lambing has gone for about a month. It's easy to neglect late lambers.
- Watch for sleepy sickness, especially in ewes that look to be carrying multiples and close to lambing.
- Check daily for cast ewes.
- Ewes may start to grow dags with lush high-protein feed. Blowfly should be little risk at this time of year.

Feeding/Pasture management

- Make sure pasture growth and feed reserves are building up ahead of the needs of stock.
- Use supplements wisely.
- Consider some nitrogen to boost feed for next month if the soil temperature is above 6°C.
- Plan and implement your soil and pasture testing programme.

Animal health

- Update animal health records.
- Make sure lambs on their mothers are growing well. If they are not, the ewes need more feed.
- Docking and castration should be done before lambs reach 3-4 weeks old.

General

- Check electric fences regularly.
- Check farm water supply regularly.
- Check financial budget and cashflow.
- Finalise cashbook.
- Pay accounts monthly.

- Contact your banker/accountant to check GST payments.
- Estimate your cash position and bank requirements for payment of creditors.

September

Livestock

- Lambing in full swing (South Island).
- Identify ewes into groups by lambing date to control feeding.
- Fully feed lactating ewes, especially those with multiples.
- Concentrate on saving every lamb. Watch out for the boredom that creeps in after lambing has gone for about a month. It's easy to neglect late lambers.
- Watch for sleepy sickness, especially in ewes that look to be carrying multiples and close to lambing.
- Check daily for cast ewes.
- Ewes may start to grow dags with lush high-protein feed. Blowfly should be little risk but be on guard if the weather gets hot and humid.

Feeding/Pasture management

- Make sure pasture growth and feed reserves are building up ahead of the needs of stock.
- Use supplements wisely.
- Consider some nitrogen to boost feed for next month if soil temperature is above 6°C.

Animal health

- Update animal health records.
- Make sure lambs on their mothers are growing well. If they are not, ewes need more feed.
- Docking and castration should be done before lambs reach 4 weeks old.
- Watch for blowfly on healing tail wounds.

General

- Check electric fences regularly.

- Check farm water supply regularly.
- Check financial budget and cashflow.
- Finalise cashbook.
- Contact your banker/accountant to check GST payments.
- Estimate cash position and bank requirements for payment of creditors.
- Pay accounts monthly.

October

Livestock

- Hoggets should be shorn (North Island).
- Consider dipping shorn hoggets (or applying pour-on) 2-3 weeks after shearing, or leave until whole flock is dipped after the main December shearing.
- Prepare wool carefully – it's always worth the effort.
- Do not dip for at least 6 weeks before shearing.
- Have good feed and shelter for sheep immediately after shearing.
- Keep all lactating ewes on the best feed to maintain lamb growth.
- Give the very best feed to ewes suckling multiples.
- Watch for infections and blowfly attacks on stumps where lambs' tails have dropped off.
- Check for footrot and foot scald as pastures become wet and lush.

Feeding/Pasture management

- Make sure pasture growth and feed reserves are building up ahead of the needs of stock.
- Feed supply should now be ahead of feed demand.
- Consider some nitrogen to boost feed for next month.
- Identify pastures to be taken out of the grazing round for early hay or silage if it is a true surplus.

Animal health

- Update animal health records.
- Check growth and health of any young stock on the farm.

- Ewes may be getting dirty on lush green feed but there may be little point in drenching them.
- If lambs are scouring badly, discuss this with your vet after doing a faecal egg count. It may be the high-protein lush spring feed.

General

- Consider the strategic use of nitrogen at 40-60kg/ha to increase silage production and fully feed the sheep.
- Apply any fertiliser needed to boost total farm fertility levels, as recommended by your consultant.
- Apply maintenance fertiliser.
- Check electric fences regularly.
- Check farm water supply regularly.
- Check that farm bikes are being maintained.
- Check financial budget and cashflow.
- Pay accounts monthly.
- Contact your banker/accountant to check GST payments.
- Estimate cash position and bank requirements for payment of creditors.
- Finalise cashbook.

November

Livestock

- Ewes suckling fast-growing lambs will be losing condition. Give these ewes the best pasture on the farm.
- The lambs, too, will be eating a considerable amount of pasture.
- Make regular assessments of feed on the farm and start saving feed on clean pastures for lambs after weaning.
- Give priority to feeding stock well, rather than closing up pastures to make supplements like hay or silage.
- Lambs born very early (June-July) will be ready for the premium early market.
- Shear rams if the weather gets warm and there is a blowfly risk.

Feeding/Pasture management

- Make sure pasture growth and feed reserves are building up ahead of the needs of stock.
- Consider some nitrogen to boost feed for next month.
- Identify pastures to be taken out of the grazing round for early silage.

Animal health

- Talk to the vet about an appropriate drenching programme for early-weaned lambs.
- Check on the health of the rams, especially their feet. They are prone to foot problems on lush feed.

General

- Consider the strategic use of nitrogen to push feed forward into summer.
- Lime can be applied any time from now until late April.
- Check electric fences regularly.
- Check farm water supply regularly.
- Review annual accounts.
- Review cashflow forecasts with bank.
- Pay accounts monthly.
- Estimate cash position and bank requirements for payment of creditors.
- Finalise cashbook.

December

Livestock

- Shear all ewes and lambs if on annual shearing.
- Prepare wool carefully – it's always worth the effort.
- Do not dip or apply pour-on in the 6 weeks before shearing.
- Dip the flock or apply pour-on 2-3 weeks after shearing.
- Have good feed and shelter available for sheep immediately after shearing.
- Wean lambs, give them a drench, and put on clean pastures with good feed.

- Do not disturb them until they settle down.
- Check them regularly for blowfly.
- Check there is plenty of water for weaned lambs, especially when it is hot.
- Put lambs in paddocks with access to shade.
- Newly weaned ewes can go on short feed (with plenty of water) until their milk supply dries up. Then they should be fed well to regain lost body weight for their next mating.
- Ewes that have reared multiples need special feeding.
- Start to sort out which ewes to retain in the flock. Get rid of any unproductive sheep that you don't want to put to the ram.

Feeding/Pasture management

- Make sure pasture growth and feed reserves are building up ahead of the needs of stock.
- Plan ahead to have good green feed available through the early part of summer to keep lambs growing.
- Identify pastures to be taken out of the grazing round for silage or hay.

Animal health

- The replacement ewe lambs are the top priority – check their growth regularly, especially if you want to mate them as hoggets.
- Do not drench sheep unless it is justified by a faecal egg count.
- Check regularly for blowfly.
- Check ram health regularly.
- Discuss with your vet the testing of weaned lambs (blood or liver samples) for cobalt and selenium.

General

- Consider the strategic use of nitrogen, provided there is adequate soil moisture and the pasture is still growing.
- Apply lime if required.

- Check electric fences regularly.
- Check farm water supply regularly.
- Check farm bikes are being maintained.
- Check financial budget and cashflow.
- Pay accounts monthly.
- Contact your banker/accountant to check GST payments.
- Estimate cash position and bank requirements for payment of creditors.
- Finalise cashbook.

January

Livestock

- Maintain a good water supply for all stock, especially during hot days.
- Try to provide paddocks with shade.
- Regularly check the growth of lambs. Provide them with the best green feed on the farm.
- Make sure the ewes are gaining weight.
- All cull ewes should be off the farm to save feed.
- Make sure any entire ram lambs not intended as sires are culled.

Feeding/Pasture management

- Make sure pasture growth and feed reserves are building up ahead of the needs of stock.
- Feed quality will be falling rapidly as the weather warms up.
- Identify pastures to be taken out of the grazing round for silage or hay.
- If pasture gets out of control (long and stalky), graze it off with cattle or use a topper. Leaving dead litter may encourage facial eczema.

Animal health

- Start facial eczema protection. Check that the doses of zinc delivered are accurate.
- Update animal health records.
- Check growth of lambs/hoggets and do a faecal egg count

if they are scouring and not thriving.
- Check rams are in good health. Get them all vet checked.

General
- Consider the strategic use of nitrogen, provided there is adequate soil moisture or if you have irrigation.
- Check electric fences regularly.
- Check farm water supply regularly, especially when the weather gets hot.
- Check financial budget and cashflow.
- Pay accounts monthly.
- Estimate cash position and bank requirements for payment of creditors.
- Finalise cashbook.

February

Livestock
- Ewes should be gaining weight.
- Rams should be in top condition now too.
- Crutch ewes in preparation for main mating (North Island) in March.
- Isolate rams from the sound, sight and smell of ewes.
- Join the ram with ewes in top condition for some very early lambs (North Island).
- Sell surplus hoggets if feed gets short.

Feeding/Pasture management
- Make sure pasture growth and feed reserves are building up ahead of the needs of stock.
- Feed quality will be falling rapidly.
- If pasture gets out of control (long and stalky), graze it off with cattle or use a topper. Leaving dead litter may encourage facial eczema.

Animal health
- Continue facial eczema protection.
- Check health and growth rate of hoggets.

General

- It's probably too dry and hot to apply nitrogen unless irrigation is available.
- Lime and maintenance fertiliser may be applied if required.
- Check electric fences regularly.
- Check farm water supply regularly.
- Check farm bikes are being maintained.
- Check financial budget and cashflow.
- Pay accounts monthly.
- Contact your banker/accountant to check GST payments.
- Estimate cash position and bank requirements for payment of creditors.
- Finalise cashbook.

March

Livestock

- Join rams to the main flock. Joining ewes and rams that have been isolated should stimulate ovulation and hence multiple births.
- Ewes should be in rising condition when joined to get a 'flushing' effect.
- If mating hoggets, join them 2 weeks earlier. (Hoggets must be a minimum of 40kg for joining.)
- Ewes may be shorn before mating if twice-yearly shearing.
- Take care in preparing the wool.
- Stock need water and especially shade.

Feeding/Pasture management

- Make sure pasture growth and feed reserves are building up ahead of the needs of stock.
- Feed quality will be falling.
- If pasture gets out of control (long and stalky), graze it off with cattle or use a topper. Leaving dead litter may encourage facial eczema.

Animal health

- Continue facial eczema protection.
- Drench hoggets on the basis of a faecal egg count. Discuss drenching programme with your vet.
- Watch for flystrike.
- Discuss with your vet the checking of selenium status in ewes by blood tests.

General

- Consider the strategic use of nitrogen
- Carry out soil tests and review your fertiliser programme.
- Apply any autumn fertiliser needed.
- Check electric fences regularly.
- Check farm water supply regularly.
- Check financial budget and cashflow.
- Pay accounts monthly.
- Estimate cash position and bank requirements for payment of creditors.
- Finalise cashbook.
- If 31 March is balance date you must lodge your books with an accountant by 1 May.

April

Livestock

- Rams should be working (North Island). Check them regularly to make sure they are in good health and doing the job.
- Use mating harnesses to check the cycling patterns of ewes.
- Use a different breed of ram at the end of mating as a 'follow-up' ram.
- Ewes should not lose weight or condition during joining.
- Cull any unwanted rams after mating.

Feeding/Pasture management

- Ideally there will have been some autumn rains to produce some green feed. If it's still dry, ewes will be losing condition – give them the best feed available.
- If it's very dry, consider feeding supplements to maintain weight.
- Start any pasture renewal needed.

Animal health

- Continue facial eczema zinc treatment into May
- Check growth and health of hoggets.
- Update animal health records.
- Check rams when they are finished mating.

General

- Consider the strategic use of nitrogen at 30-50kg/ha to build up late autumn/early winter feed.
- Check electric fences regularly.
- Check farm water supply regularly.
- Check financial budget and cashflow.
- Pay accounts monthly.
- Contact your banker/accountant to check GST payments.
- Estimate cash position and bank requirements for payment of creditors.
- Finalise cashbook.

May

Livestock

- Check that rams are working and ewes are cycling (South Island). Use mating harnesses to check cycling patterns of ewes.
- Use a different breed of ram at the end of mating as a 'follow-up' ram.
- Ewes should not lose weight or condition during joining.
- Cull any unwanted rams after mating.

Feeding/Pasture management

- Ideally there will have been some autumn rains to produce some green feed. If it's still dry, ewes will be losing condition – give them the best feed available.
- If it's very dry, consider feeding supplements to maintain weight.
- Start any pasture renewal needed.

Animal health

- Check growth and health of hoggets.

- Update animal health records.
- Check rams when they are finished mating.

General

- Consider the strategic use of nitrogen as for April.
- Check electric fences regularly.
- Check farm water supply regularly.
- Check financial budget and cashflow.
- Pay accounts monthly.
- Contact your banker/accountant to check GST payments.
- Etimate cash position and bank requirements for payment of creditors.
- Finalise cashbook.
- If 31 May is your balance date you must lodge your books with an accountant by 1 July.

9. How Much Sheep Eat

Stocking rates

The maximum number of sheep you should farm is based on the number of sheep your pasture can feed over winter when pasture growth is minimal. This is your farm's winter *carrying capacity*.

Your *stocking rate* depends on how much pasture dry matter (PDM – the weight of pasture less the water content) you can grow, *and* how much supplementary feed (hay and silage) you can make on the farm. Table 2 gives a rough idea of how pasture length relates to the amount of pasture dry matter (PDM) per hectare.

The table shows the minimum pasture length required by various types of sheep. You can use it to help plan your rotation. For example if you have 80 ewes after weaning in summer on good pasture they will need pasture at least 1-2cm long. You may have a two-hectare paddock with this length of pasture. The table shows that total PDM will be about 1000kg/ha, i.e. there is 2000kg dry matter in the paddock. The mob needs 80kg DM/day so the paddock should last 25 days (2000 divided by 80).

You can carry more stock again if you have or are prepared to buy in supplementary feed (hay, grain, meal or concentrate pellets) in addition to pasture over winter.

Table 2. Recommended minimum pasture length and pasture dry matter (PDM) quantities

Sheep type	Pasture length (cm)	Pasture DM (kg/ha)	Feed intake (kg DM/ day)	Growth rate (grams/day)
EWES				
Mid-pregnancy (winter)	1-2	400-500	1.0	Maintenance
6 weeks pre-lamb	2-3	600-800	1.3	60-80
Ewes with lambs (spring)	4-5	1400-1600	1.8	180-200 (lambs)
Post weaning (summer)	1-2	900-1000	1.0	Maintenance
Mating (autumn)	2-3	1200-1400	1.4	120-150
LAMBS				
Weaned (spring)	3-4	1200-1400	0.8	160-200
(summer)	2-3	1400	1.0	130-150
(autumn)	2-3	1200	1.2	80-100
(winter-spring)	3	1100	1.2	100-120
1-YEAR-OLDS	2-3	1400	1.3	60-80

Adapted from tables in the *Code of Recommendations and Medium Standards for the Welfare of Sheep* (1996).

On hill-country farms, stocking rates are around 4-6 ewes/ha, and good, well-fertilised lowland farms can usually carry about 11 ewes/ha over winter.

The concept of the 'livestock unit' (LSU) or 'ewe equivalent' is often used to work out the total carrying capacity of a farm at 30 June in any particular year. An LSU is expressed as the annual feed requirement for a 55kg ewe rearing a single lamb. Other livestock are expressed as a proportion of LSU or as multiples of it.

In feed dry matter terms one LSU needs 520kg of good-quality pasture per year to meet its needs. Tables 3 and 4 show the PDM requirements of sheep of various ages.

Table 3. Recommended daily pasture dry matter (PDM) requirements for lambs and one-year-old sheep

For maintenance plus liveweight gain per day (in kg)				
Liveweight	Maintenance	50g/day gain	100g/day gain	200g/day gain
20kg	0.6	0.7	0.9	1.1
30kg	0.8	1.0	1.2	1.5
40kg	1.0	1.2	1.4	1.9

Table 4. Daily PDM requirements for ewes (in kg)

Liveweight	Maintenance	Late pregnancy	Lactation
40kg	0.74	1.1	2.2
45kg	0.80	1.1	2.3
50kg	0.86	1.2	2.4
55kg	0.92	1.3	2.5
60kg	0.99	1.3	2.6
70kg	1.12	1.4	2.8

Notes

- Pasture gives about 11 megajoules of metabolisable energy (MJME) per kilogram of dry matter (DM).
- Average pit silage gives 8.5 MJME/kg.
- Average hay gives 7.5 MJME/kg.
- Good-quality hay and silage give about 10 MJME/kg.

What sheep want vs what they need

Sheep will keep on eating as much feed as they can. The trick is to balance their wants with their needs.

Their nutritional needs depend on their age, gender, weight, whether or not they're pregnant and the season of year. It takes experience to know how much pasture to feed, and how and when to supplement it with hay, silage, concentrates and/or grain.

It is wise to get training or advice at the outset from a consultant or an experienced farmer.

Getting the feed supply 'right' may seem complicated, and you may not want to bother. You can just move the sheep onto fresh pasture when you think they need it, or when they start verbally complaining when they see you. But if you're interested in optimum production and profit it is generally far better to feed them the appropriate amount.

A 'maintenance' diet provides the nutrients needed to maintain the animal's body weight, regardless of its body condition. On top of this there's the feed needed for 'production', i.e. the feed needed to provide nutrients for things like movement, body growth, wool growth, pregnancy and milk production.

Sheep needs vary

Pregnant ewes

Pregnant ewes of all ages should be well fed for a few weeks after mating, and then again in the last six to eight weeks of pregnancy. During this latter time their appetites may drop off, as a result of hormones and the increasing size of the uterus, so be aware of this and provide a high-quality diet or some concentrates.

This goes against past practice, when ewes were made to clean up pastures after mating and kept on a low plane of nutrition before lambing to avoid difficult births. Today's sheep are more productive and need better care right through pregnancy. (See 'pregnancy diagnosis' in Chapter 12.)

Pregnant two-tooths need special attention as they approach their first lambing, and any ewes scanned as carrying twins should be put on the best feed available, with good-quality supplements when needed.

Ewes scanned as carrying singles require a little less feed than twin-carrying ewes throughout pregnancy. They can be run on slightly less pasture to prevent their lambs becoming too big and causing difficulties at birth. But be careful not to cut back their feed too much or you'll reduce their milk production or trigger metabolic diseases such as sleepy sickness.

Any ewes scanned as empty should be disposed of, or put on short feed. But note that because scanning is not perfectly accurate, a small proportion of 'empty' ewes may in fact be in lamb – probably very late.

Lactating ewes

The ewe's milk supply is the biggest factor influencing lamb growth from birth to weaning. Its most critical importance is in the first four to six weeks of life, before the lamb starts to eat substantial amounts of pasture.

Lactation pulls more nutrients from a ewe than pregnancy. Lambs, unlike calves, suckle many times a day, especially when they are very young. As you'd expect, ewes suckling multiples produce more milk than those suckling singles, as the udder is emptied and stimulated more often.

This then triggers the ewe to eat more, and if she's not well fed she'll milk condition off her back – i.e. she will lose weight as her body condition is sacrificed to benefit her lambs.

Ewes suckling lambs must have the very best feed available to keep their milk production going. You cannot overfeed ewes suckling lambs, especially those rearing multiples.

After about four weeks old the lamb will start to eat pasture in competition with its mother.

Ewes after weaning lambs

On weaning day, reduce the ewe's feed to reduce her milk supply. Watch those that have been suckling two or more lambs as they will have larger udders and are more prone to mastitis after weaning. Ewes that have been poor milkers will have dried off before the lambs are weaned, so they won't require food restrictions.

Do not reduce the water supply to weaned ewes.

About four weeks after weaning, provide good feed to build up the ewes' body condition again before mating and winter.

Do not cull lean ewes after weaning, as you may cull the ones that have milked well and earned their keep!

Ewes before mating

Ewes should be fed on a steadily increasing ration (a rising plane of nutrition) for at least three weeks before the ram goes out. This is called 'flushing': it's a tried and true trick of sheep farmers because it causes multi-ovulation, i.e. more eggs are released from the ewe's ovaries.

Pasture has to be saved for flushing, and this is often difficult during a dry autumn. Some farmers grow a brassica crop especially for the purpose.

Newly shorn ewes

Shearing increases the appetite of sheep for several weeks afterwards – by up to 30% in summer and up to 70% in winter. This is caused not just by a response to cold but by an increase in the sheep's metabolism. Its engine is revving faster! So put newly shorn sheep on good feed for up to six weeks after shearing, and make sure they have good shelter in rough weather.

Hoggets

Feed for ewe hoggets and weaned ewe lambs should never be restricted, especially if you want to put them to the ram. They are very important for the future of your flock and need special care.

Sheep destined to lamb as hoggets will need priority feed until their lambs are weaned, as they're expected to grow and reach mature size as well as nurture a foetus and produce milk. It's a tall order on many farms, and it's easy to forget the extra requirements of growing and pregnant hoggets when calculating feed needs and carrying capacity.

Hogget liveweight at 14 months is an important measure. It has a major effect on lifetime productivity expressed in future growth, body size, fertility and wool production.

Body condition scoring

It is important to know your sheep's 'body condition', as this is a good measure of their health and well-being. 'Liveweights' (simply the weight of the sheep) can be misleading unless you know if the sheep has a small or large frame.

Their wool cover can make it difficult to determine their body condition. You must put your hand on the sheep's loin – over the backbone between the last rib and the pelvis – and feel through the wool for fat and muscle cover. (You're checking out those loin chops!)

The body condition scoring system allows you to assess body condition regardless of body size. The scale ranges from completely emaciated (0) to extremely obese (5).

Body condition score

Grade 5: The bones of the spine are undetectable, even with firm pressure, and there is a gutter along the backbone over the spine. The horizontal processes cannot be felt. The muscle is full with a very thick fat covering, often with accompanying heavy deposits of fat in the rump area. Sheep in this condition are overfat and require careful feed management.

Grade 4: The bones of the spine are difficult to find below the fat-covered muscle areas. The ends of the horizontal processes cannot be felt readily. The muscle is full and thickly covered by fat. Sheep in this condition are close to becoming fat.

Grade 3: The bones of the spine can be felt with gentle pressure, and they feel smooth and well covered. The muscle is full and rounded and there is a moderate depth of fat cover. Sheep with this grading are in good body condition.

Grade 2: The bones of the spine are prominent but smooth. There is muscle present but little fat cover.

Grade 1: The vertical spine and horizontal processes are prominent and sharp. There is no fat cover. Muscle can be felt but it is wasted. These sheep are in poor condition and require preferential treatment.

Grade 0: No muscle or fat can be felt between skin and bone. Sheep in this condition are emaciated and very weak and may have to be destroyed unless preferential treatment can be given rapidly. They are unfit to travel, unsuitable for human consumption and should not be sent to a meat processing plant.

10. What Sheep Eat

Pasture

Pasture is a very variable feed. It varies in both quantity and quality from week to week – and even sometimes from day to day depending on the weather.

Good-quality pasture is generally the best and cheapest feed for sheep. Surplus spring feed can be made into hay or silage and stored for supplementary feeding in winter, and surplus autumn pasture can be saved so that it also can be fed in winter.

Pasture that has been left ungrazed gets stalky and rank. It may also be too long for sheep, and be more suitable for cattle or horses. Sheep graze much closer to the ground than cattle and they prefer shortish pasture and can cope if it is as low as 2cm long. About 5cm long is ideal.

Long-term overgrazing with sheep will produce a very dense lawn-type sward. Rabbits love this kind of pasture and the two animals together have been the cause of much land degradation. The ideal combination is mixed grazing, with sheep and cattle together, or sheep and horses/ponies. The bigger animals eat the long pasture, leaving the shorter leafy material for the sheep.

As a general rule, it's a good sign to see ruminants like sheep standing or lying down chewing their cud. Although it doesn't necessarily mean the pasture is of good quality, it generally means they are getting the quantity they need.

Sheep are like kids at a party – eating the best stuff first and leaving the less tasty food until last or just avoiding it.

Fungal toxins (mycotoxins) can grow in dead litter at the base of the sward in late summer and autumn, and these can cause health problems. The most serious are the toxins that cause facial eczema, and the lesser-known fungal toxin zearalenone that causes reduced fertility. This is a good reason for controlling surplus pasture to avoid build-up of dead litter.

Select good pasture species

Make sure your pastures have productive species of grass and clover, and that there are few weeds and patches of bare ground. If your pasture is poor, talk to a pasture consultant about a pasture renewal programme for autumn.

Check out the latest recommendations for the species and cultivars (varieties) for your soil type and climate. In areas where ryegrass staggers is a problem, choose endophyte-free or endophyte-safe cultivars of ryegrass. (Endophytes are fungi that can grow inside some types of ryegrass.)

When sowing new pasture, include:

- One grass (or maybe a perennial ryegrass plus a hybrid short-term or long-term ryegrass).
- At least two white clovers, one for grazing and one smaller-leaf type for nitrogen fixing and colonising the pasture bottom. Clover is a highly nutritious feed for sheep: rich in protein, energy and minerals. Clover can be encouraged by ensuring soil fertility is kept high, and by not allowing grasses to grow tall to overshadow the clover. Clover root nodules enrich pastures by 'fixing' nitrogen from the air – free of charge!
- Perhaps a grazing herb such as chicory or plantain for improved pasture quality and variety in the sheep's diet.
- Ideal sheep pasture contains about 30% clover and 70% grasses (and looks as if it's mostly clover!).

Soil tests and fertiliser application

To grow good pasture, it is very important to optimise soil fertility. To this end it's a good ideas to have soil tests carried out at regular intervals (say every two or three years). Arrange for the results to be sent to you with interpretative comment by an expert who will recommend the appropriate fertilisers for good pasture growth and minerals for good stock health, and when and how often to apply them.

Getting good advice on soil type and following through on the recommendations may seem expensive, but it's an important basic step toward good profitable farming.

Dealing with surplus pasture

On small farms, spring pasture can easily grow far faster than the sheep can eat it, so it's important to have a number of options for controlling the surplus. Consider some of these options:

- Use cattle to eat the longer rougher pasture, and graze sheep on the nutritious regrowth.
- Buy more sheep to try to control the surplus and sell them when it's gone. These could be dry ewes or wethers, for which quality of feed is not as critical as with ewes and lambs.
- Borrow some stock such as mature cows to eat off the surplus – a quick 'in and out' strategy.
- Preferentially graze the faster-growing pastures, keeping the slower-growing pastures in reserve.
- Make the surplus pasture into silage or hay. Sheep will enjoy the regrowth after cutting.

Making silage, baleage and hay

In spring, surplus pasture can be made into hay or silage. See a contractor early in spring to make sure he or she can do the job for you – and that you are not too far down the list of small blocks to be serviced! Contractors prefer large paddocks with few gates.

Your contractor will also advise on the suitability of your pasture for hay or silage and the right time to cut it.

Silage

Silage is made from young grass high in protein, cut when it starts to show seedheads (15% seedheads). It is then wilted and stored in airtight containers (silos, bunkers or plastic wrap) so that the juices become acidic and preserve the nutrients – as in pickling. It usually has dry matter content (DM) of 25%. Good silage is olive green in colour, with a sweet and fruity smell. If it's brown, mouldy or smells bad, don't give it to any livestock.

Baleage

Baleage is made from slightly more mature pasture, and wrapped in plastic wrap after wilting. It has 40% DM content but lower feed value (energy and protein) than silage.

Hay and straw

Hay is made from more mature pasture or lucerne. It is thoroughly dried before being baled and stored under cover. It has very high 85% DM content but it's low in protein and high in fibre. Good hay is leafy and sweet smelling, not mouldy. Sheep like leafy hay that is not over-mature when cut. Good clover content helps make good-quality hay.

Lucerne hay has the best feed value if has been dried correctly. It makes good sheep feed.

'Stubble hay' is the pickings from a paddock that has had a grain crop harvested.

Straw consists of the stalks left after harvesting grain. Oat straw has more nutritive value than barley straw, which in turn has more than wheat straw, but generally all have low nutritive value and sheep will not usually eat straw unless they are very hungry.

Rotational grazing

Rationing pasture (sometimes called strip grazing) involves grazing relatively small areas of pasture at a high stocking density for a short time, and moving stock frequently. Rotational grazing has to be planned well in advance.

Compare this with conventional 'set-stocking' systems, where sheep stay in the same paddock for long periods.

Sheep in rotational grazing systems can be moved daily or every 2 or 3 days depending on how many sheep there are (and how much time you have to move the fence!). Usually electric fences are used to define the area to be grazed.

This method of grazing can help to reduce internal parasite numbers as sheep are generally moved onto pasture that has relatively few worm eggs because it has not been grazed by sheep for a while. It also helps make effective use of pasture. Sheep graze quickly and effectively before pasture is wasted by trampling or becomes fouled by dung.

Sheep soon learn to graze during the early part of the move while the pasture is still fresh, and they may rest until the next move.

For maximum efficiency, use a back fence to contain ewes in each block to allow rapid pasture regrowth behind them. You may then have to provide water in portable troughs, which are readily available to all stock.

Supplementary feeding in winter

Winter is the time when feed is often short, and if there is a chance that there will be insufficient pasture, it's important to have supplements such as hay, silage and baleage on hand.

Other common supplements include concentrate pellets and grain. Ewes pregnant with two or more lambs will probably require such concentrates or grain as well as hay in order to provide the energy and protein they need. Sheep in poor body condition also benefit from concentrate or grain feed, especially in winter, as hay and silage don't provide enough nutrients.

How much hay should I feed?

Hay is the most common supplementary feed – and the easiest to feed. Good sheep hay is sweet smelling, leafy and high in clover content. Avoid mouldy hay at all costs – and hay that is dusty or discoloured. Mould or fungus make the feed unpalatable, but if sheep are forced to eat it they may lose weight and/or develop diarrhoea (scours). Also avoid coarse, stalky over-mature hay that has lost most of the grass seeds and/or contains weeds such as thistles and ragwort.

When it comes to feeding hay there is a useful rule of thumb. Most grazing animals require the equivalent of about 2% of their body weight in good-quality hay daily. This means a 50kg sheep requires at least 1kg or a twentieth of a small conventional bale of good-quality meadow hay daily.

Remember that some types of livestock require more than this basic maintenance ration.

If the hay is of poor quality or if some is lost by trampling, more should be given accordingly.

Introduce new feed gradually

Sheep have to learn to like any new feed, and they learn from other sheep and especially their mothers.

Any new foods, including grain, concentrates, silage and baleage, should at first be offered in very small daily amounts. The amount offered should gradually be increased to the full ration over a period of one to three weeks. This is called 'preconditioning'.

Feed must be spread out so that shy feeders get a chance to eat and the greedy individuals don't gorge. Sheep have a well-defined pecking order: you will see who's boss at feed time. If the competition is too great you may have to feed the shy sheep in a separate group.

Feeding concentrates or grain

Sheep have to acquire a taste for concentrates or grain, and lambs learn from their mothers. Such feeds should be fed with roughage such as pasture, stubble hay or straw, and not as the sole diet.

Only about 50g grain should be offered per sheep for each of the first 10 days. Spread it out in a long line to give them all a chance to eat it. A sheep that eats more grain or silage than it is accustomed to may develop rumen acidosis or grain overload, which is often fatal. Fifty grams is enough for most sheep, though you may wish to build up to 100g a day for ewes with multiple lambs.

Oats are probably less likely to cause problems than wheat, which requires a more prolonged period of preconditioning.

Hazardous foods

Apples, potatoes, corn waste and other chunky foods can be hazardous because hungry stock may choke. A sheep that is choking coughs and splutters and may shake its head. It will carry its head low and pushed forward, and can die of asphyxia in minutes. Try clearing the blockage by giving the sheep a very sudden sharp squeeze around its chest, with its head stretched forward and neck extended.

Too much of any new food too soon can result in bloat, rumen overload or some other type of indigestion.

Some vegetable waste is simply not sufficiently nutritious on its own and should be fed only in conjunction with more nutritious feeds.

Water

Water is vital for sheep. It is absolutely not true that sheep get enough water from pasture to meet their needs.

It *is* true that Merino sheep in very dry conditions (such as the Australian outback) can survive with moisture from the morning dew, but they are in survival mode and slowly losing weight day by day. They have also adapted to these conditions over a very long time.

Clean, fresh, potable water must always be available to help keep the sheep's digestive system healthy, even in winter. This is particularly important if dry foods such as hay or concentrates are being fed. The average mature sheep is said to need four litres of water per day, and a lamb about one litre per day. In very hot conditions these amounts increase by about 20%.

Ewes suckling lambs, especially multiples, and particularly when it is warm, will need much more than four litres/head/day.

Water troughs should be cleaned regularly. It is not uncommon for vermin or birds to drown in troughs, and lambs at the playful stage can fall in and drown. Throw some large rocks into the trough or a plank of wood, or put mesh across the top to prevent drownings.

Generally, stagnant water sources such as farm dams and ponds are not good for stock, because of dung contamination and the risk of sheep getting bogged in mud at the edges. If you do use a dam as a water supply it may need to be cleaned out periodically by a contractor using a digger.

11. Identification, Weighing and Record-keeping

Ear tags

Whatever the size of your flock, each animal should be marked or tagged with a unique identification number (ID). A unique ID is best made up of an individual number, and the last two digits of the year in which the sheep was born. For example a ewe with an ID of 35/02 would be animal number 35 born in 2002. No other sheep in the flock should have that number. You can have sheep with number 35 each year, but they'll have a different year born, e.g. 35/03 or 35/04.

Small brass ear tags are best for the main ID. These are best put in a day or two after birth, making sure you put them in the right place. Tag stockists (e.g. stock firms) will print the numbers you want on the tags. It is a good idea to have your farm or stud name on the tag too.

Punch the tag in the top of the ear, about a third of the way along from the head toward the tip, leaving room for the ear to grow and the tag to remain in a readable position. If you put a brass tag too close to the head it will become grown into the skin folds. Any heavy tag put beyond the halfway point along the ear will cause the ear to droop.

Sheep's ears are very sensitive indeed – they hate having them pulled, let alone having holes punched in them! Handle ears gently. Insert the tags where you can read them close up without having to touch the ears (if you can).

When tagging keep your hands, the tags and the equipment

spotlessly clean and disinfect frequently. Otherwise the sheep's ear will become infected and painful. After tagging check for bleeding and a few days later check for infection – you may have to remove the tag to clear up infections, which are very painful for the sheep.

After weaning, you can put a plastic tag in the ear. These are easier to read than small brass tags. There is a wide choice of excellent plastic tags available these days – choose a type that meets your needs.

If you use a plastic tag as well as a brass tag, put the same number on both if you can, to prevent problems later trying to sort out which sheep is which in the records.

Don't reuse old tags – this leads to duplication of numbers and confused records.

Some farmers with large non-recorded commercial flocks 'age mark' their sheep by taking notches out of the edge of the ear or bits off the tip, to denote the year born. There should be no need to do this if you have a good tagging system.

Despite manufacturers' claims, the numbers on some plastic tags fade with age, so renew numbers with an indelible marker before they disappear completely.

You may at some time need a temporary ID. This is most easily done with a spray paint or raddle on the wool. You can even write large numbers on the side of the sheep, but make sure you use an approved product that will scour out of the wool before processing – and use as little as possible. Place an aerosol can nozzle close to the wool and give only a very short burst of spray to keep the colour localised.

Why weigh?

A set of scales (i.e. a weighing platform) is essential if you are serious about making sheep farming profitable. They will give you accurate body weight measurements, which are useful in a number of areas. Share them with neighbours or borrow them from your vet clinic.

Ensuring correct dosing

Accurate animal weights are needed for effective anthelmintic dosing. The effective dose is based on the body weight of the sheep. Underdosing is ineffective and encourages drench resistance. Even the heaviest sheep in the group must get its full dose.

Overdosing can also be hazardous, especially if mineral supplements such as selenium or copper have been included in the drench. An accidental overdose with even two or three times the recommended dose of selenium or copper can be fatal.

Which is the heaviest?

Scales are essential when selecting replacement sheep for your flock, as you will probably want to keep the heaviest hoggets and two-tooths.

Scales are also essential when picking lambs for the meatworks, to make sure you maximise returns. If you use scales then buyers who claim to be able to estimate weight by eye cannot underestimate the weight of your sheep!

Weighing tapes that go around the girth of the sheep can give a rough estimate of weight, but they are not totally accurate.

Wool weight

Smaller scales are needed for weighing wool, although some electronic weighing platforms may be accurate enough even for small amounts of wool. It's important to know how much wool individual sheep produce if you want to select the best wool producers for your flock.

The wool weight of hoggets is particularly useful, as it predicts their potential lifetime wool production. The fleece weight of older ewes is affected by the number of lambs reared and other variables.

Keeping records

Don't rely on your memory. You can easily forget important details about your sheep, especially over a number of years. Good records are essential to making good decisions about which sheep to keep and which to cull.

You may choose to join an official sheep-recording scheme (contact Sheep Improvement Ltd). If your flock is too small to bother with the demands of an official recording scheme, then design something for yourself.

The basis of this can be a record card for each ewe, or a spreadsheet. This should record the number of lambs the ewe produces each year, their gender, weaning weight (if you have scales), and what happened to them. From this you'll readily be able to identify the good old ewes that produce heavy lambs every season. These can be the mothers of your replacement ewes and rams. It's as simple as that.

You can record more details if you wish, such as mating dates, lambing dates, health records, fleece weight and quality, and so on.

But remember, unless you enjoy collecting records, collect only the information you will use for selection and breeding decisions to improve the flock's profit.

It's a good idea to keep a duplicate set of records somewhere. Records get lost, they can be burned in house fires, some have been put through the washing machine and some have even been eaten!

Pencils write on wet paper better than ballpoint pens, and a clipboard inside a plastic bag is good when working in yards. In wet weather use a piece of card to rest your hand on while filling in records. It keeps the rest of the sheet dry. An old towel around your neck to dry hands is good idea too. The ideal, of course, is waterproof paper, which you can buy in some notebooks designed for surveyors.

12. Mating

Ewes are stimulated to start cycling and come on heat (oestrus) by the gradual shortening of days and lengthening of nights in autumn.

This change also stimulates rams into a kind of 'rut', when they start to smell strongly (like billy goats), and areas of bare skin along the belly inside the front and hind legs turn a pinkish colour. They're 'in the pink'!

Vital statistics

- A ewe comes on heat every 17 days (with a range of 14 to 21 days).
- Each heat lasts for about 24 hours (with a wide range of four to 72 hours).
- Sheep reach puberty (sexual maturity) between six and 10 months of age.
- Pregnancy length (gestation) ranges from 140 to 150 days (about five months).

When to mate

The best time to put a ram ('join' the ram) with your ewes depends on when you want ewes to lamb. Remembering that pregnancy lasts for five months, the decision on when to start lambing can be based on:

- When the spring pasture really starts to grow on your farm.

- A preference for lambing when the weather is better.
- When extra people (such as family) are around to help.
- A preference for early fat lambs for premium prices.
- Lambing in winter to catch the premium market at Christmas.

Rams

Good active rams will 'sniff hunt' around the ewes in their flock several times a day to check for those on heat. In large paddocks the rams often form a harem, keeping a small group of ewes with them all the time.

The ram produces pheromones (a chemical found in the wool grease) that further stimulates ewes to ovulate (shed eggs). Some breeds, such as the Polled Dorset, seem to be particularly good at producing pheromones!

If you are using young rams (ram lambs), put them with older experienced ewes, and put an old experienced ram with ewe hoggets.

You can keep track of proceedings by using a ram harness with a coloured crayon on the brisket to record matings. When the ram mounts, he leaves a crayon mark on the ewe's rump. If you change the colour every 17 days (use the traffic light sequence), this will tell you:

- Which ewes are not pregnant (they are multi-coloured, having returned to the ram throughout the season).
- When they'll lamb (it's handy to know if they'll be early, middle or late season for feed budgeting)

Be aware, however, that the ram harness is not foolproof, as some rams don't leave much of a mark, and pale colours such as yellow may be faint and easily missed.

If there is any hint from recurring raddle marks that the ram has low fertility (i.e. if many ewes continue to cycle), then change your ram. You can get a ram tested before mating to see if he is producing viable sperm. If your whole breeding programme depends on one ram, then the cost is probably justified. Talk to your vet about it.

Most farmers leave the ram in with the ewes for two or three cycles (two or three months), depending on how long they want lambing to go on for.

Ewes

Oestrus ('heat') signs in ewes are not very obvious to humans. You may see ewes:

- Seeking out and sniffing rams, especially soon after the rams have gone out.
- Crouching and urinating when a ram approaches from behind.
- Standing still to be sniffed and mounted.
- Turning her head around to watch a ram's approach.
- Tail fanning – rapid shaking of her tail when the ram sniffs.

Putting hoggets to the ram

You can put hoggets to the ram at seven to eight months of age, but there are a lot of things to get right if you want to ensure success. If you put too much pressure on such immature animals they can be ruined for life.

Take note of the following points:

- They must be heavy enough – aim for minimum body weight of 40kg (use your weighing platform).
- Use an older experienced ram to mate them.
- Don't expect 100% to become pregnant; 50-60% is more usual.
- It's a good idea to scan them to find out which are pregnant so that they can be given preferential treatment. Pregnant and lactating hoggets need extra feed not just for their own growth, but also to nurture their lamb both before and after birth.
- Breeding from hoggets will have an impact on feed management for the rest of the flock. Make sure you have the feed to give them.
- Do *not* treat hoggets with Androvax (a vaccine used to increase the number of lambs born) – they have enough to do with one lamb without the extra stresses of rearing twins (or triplets).

Pregnancy diagnosis

A contractor with an ultrasonic scanning machine can determine whether a ewe is pregnant and how many lambs she is carrying. The scanner is placed on the underside of the ewe's belly while she is held in a crate in the race or turned on her haunches.

For best results, pregnancy scanning should be carried out around 100 days after the ewe has been mated.

Up to 35% of the foetuses seen on scans may be lost before lambing. This is very frustrating, and there are several possible reasons for it, including foetal re-absorption (usually with no obvious cause) and false diagnosis as a result of operator error.

However, most operators are very accurate and the procedure is not risky for the ewe or unborn lamb.

You have to decide whether the cost is likely to be justified, bearing in mind the fact that the operator may charge more per head for small numbers.

13. Lambing

Prepare early

You'll notice that ewes heavy in lamb and 'on the drop' are very serene and slow up a lot. Don't hassle them. Always move them quietly, and banish enthusiastic dogs!

Their appetite may also drop, especially if they're carrying two or more lambs. This might be partly due to their enlarged womb, but it is also a hormonal effect.

A few weeks before lambing, the pregnant ewes should be moved to a clean lambing paddock with good shelter from cold, wind and rain. A ewe will take time to find a quiet spot and prepare a birth site.

The most likely health problems at this stage are acetonaemia or sleepy sickness (see pages 142–143) and bearings (see pages 145–146).

Make sure you have all the equipment needed well ahead of time:

- Lambing lubricant
- Disinfectant
- Rubber gloves
- Feeding tube for weak lambs that can't suck
- Covers for lambs in wet cold weather – bought or homemade
- Heat lamp and box to warm chilled lambs
- Treatment for ewes with sleepy sickness
- Pessary for ewes that develop metritis (infection of the

uterus), usually after a difficult birth (consult your vet first)
- Bearing retainers (see pages 145–146) (consult your vet first)
- Antibiotics for ewes with mastitis (consult your vet first)
- An offal hole protected by a safe cover

Traditionally, Scottish shepherds always got a bottle of whisky in for lambing time – for the sheep of course!

Signs that lambing is imminent

A ewe will 'bag up' (her udder will swell) a week or two before lambing. Closer to the time her vulva will swell and there may be some mucous discharge. At this point she will separate from the flock to find a quiet place to give birth. She will prepare a 'birth site' by smelling the ground, pawing it with her front feet and going round in circles. She'll lie down and stand up again quite a lot as labour begins.

The normal birth

Fortunately nature has been delivering young animals with no human help for eons, and the great majority of farm animals give birth naturally with no problems.

Once the birth process begins, try to keep a watchful eye on events without making the ewe aware of your presence.

Womb contractions position the lamb for delivery. The ewe usually stands during this phase. Fluid and the placenta (the water bag) appear at the vulva, and the placenta may bulge or hang from the vagina. It will swell, then burst.

The ewe may smell the ground where the placental fluid has spilled. The smell of the birth fluids on the ground will tie her to that spot. It's on this 'birth site' that the maternal bonding takes place - the ewe may stay around that spot for a couple of days.

Lambs, like most farm mammals, are usually born in a diving position – with the front legs fully straightened so that their knees are alongside their muzzle, making a streamlined shape for ease of delivery. The two front feet of the lamb appear at the vulva, and behind them should be its nose.

As a general rule, once birth contractions begin, there is usually fairly rapid progress within 15 minutes. The ewe usually lies for the final few pushes. She may have two or more attempts at this, then the lamb is on the ground, shaking its head. Sometimes the ewe will stand up for the final push.

When the lamb is finally pushed out, the membranes should break and the ewe will turn around to lick the lamb and chew the birth membranes.

Normally, the lamb will shake its head and break the membrane over its nose, snorting to clear its nostrils so that it can breathe. The umbilical cord will stretch and usually break when the ewe turns around, but if it doesn't, don't break it until the lamb has started to breathe.

The afterbirth (placenta) will usually be passed soon after the lamb is born but it may take an hour or two. The ewe may start to chew it. This is normal and does her no harm.

If the lamb is born in a place that is at all dusty or muddy, spray or paint its navel and cord with tincture of iodine (iodophor spray). Try to avoid disturbing the ewe.

It is wise to repeat the iodine treatment 24 to 36 hours later if conditions are particularly wet.

Don't interfere unnecessarily

Watching a normal birth can be a wonderful and moving experience, especially for children.

However, all ewes have a strong instinct to give birth alone in a quiet, safe and sheltered place with no disturbances. It's nature's way of giving the ewe and her lambs a good chance of bonding, and giving the lamb time to find its feet, find a teat and get a good first feed of colostrum.

Good maternal bonding is vital in reducing lamb losses around birth – especially in ewes with multiple births. It is very important not to disturb ewes unnecessarily at any stage of the process, from the first signs that the birth is imminent to the stage when the youngster has a full tummy and is confidently striding alongside its mother.

Keeping disturbance to a minimum is especially important with first-time and/or nervous mums – their mothering instincts are more easily overwhelmed by events! So don't shift a ewe from the birth spot – let her move away with her lamb(s) in her own time. Shifting her earlier increases the risk of mis-mothering, as she'll have a very strong urge to go back to the original birth site because of the smells there.

When the lamb gets stuck in the birth canal

The birth can be complicated if the lamb becomes stuck in the birth canal. If you don't think you could deal with it (or recognise the early signs of a problem) then consult your veterinarian well before the start of lambing.

You need to intervene if the ewe has been down straining or if the membranes or part of the lamb have been protruding from the vulva for 10 minutes or more, with no visible progress. It may mean the lamb is in the wrong position in the vagina, i.e. its head is twisted to one side or a leg is bent back. Someone – preferably with experience and a small clean hand – needs to gently but firmly push the lamb back toward the womb, and straighten the limbs and body so that it is positioned correctly.

If you are the nominated one, lie the ewe on her side and put your knee or arm across her neck while inspecting her rear end. You'll then have to try to squeeze your hand into the vagina. You may need to get someone to hold the ewe by the back legs and pull them up gently so the weight of the uterus falls away from you. It's then much easier to get your hand in and manipulate things without damage.

The following hints may be useful:

- If you are on your own, it can help to tie some baler twine to each of the ewe's back legs and around behind your shoulders to take some of the ewe's weight.
- If there are lambs' legs in the birth canal try to determine if there is one lamb or two – don't start pulling till you have sorted that out!
- Know the difference between a back and front leg. A back leg has a hock; a front leg a knee.
- Sometimes the hind feet come first, or there are twins alongside each other, and this can be very confusing.

- When the tail comes first and the hind legs are extended forward under the body, this is a breech birth. These can be very difficult to deal with, and prompt veterinary attention is usually necessary.
- If you have to push a lamb back into the uterus, put a string or tape on a leg so you can find it again.
- It can be very difficult to reposition a lamb if the ewe is straining and the birth canal is tight and dry, so it helps to use lots of lubricant.
- Once the lamb is out, clear its nose and mouth of mucus and membranes, and if necessary hold it up by the hind legs and give it a gentle shake to clear mucus from its lungs.
- Don't break the cord until the lamb is breathing well.
- Keep the ewe down, rub the lamb or membranes on her nose and quietly disappear. Keep an eye on her and hope she gets up and smells/licks the lamb.
- If she doesn't show any interest in the lamb you have to catch her and confine her and the lamb in a pen until she accepts it and lets it suck.
- If after a few hours the lamb has not succeeded in sucking, help it by supporting its hind end (not its head) and holding it in about the right position to suck. Gently rub the base of its tail as its mother would if she were licking it, and its strong nuzzling and sucking instincts should take over.
- If the lamb is weak, you may have to squeeze some colostrum from the teat into its mouth. It may be easier to sit the ewe on her haunches to do this.

Call your vet if the ewe becomes dull a few days after lambing or if she has a smelly discharge from her vulva. She may have metritis, an infection of the uterus.

Lambing percentages

Lambing percentage is one of the best indicators of your flock's performance and profit potential. Other farmers will ask you for a figure so you'd better have it ready!

Lambing performance can be measured in the following ways:

1. Number of lambs born/100 ewes joined with the ram
2. Number of live lambs born/100 ewes joined
3. Number of lambs born/100 ewes mated
4. Number of lambs born/100 ewes lambing
5. Number of lambs docked/100 ewes joined
6. Number of lambs weaned/100 ewes mated
7. Number of lambs weaned/100 ewes lambing
8. Number of lambs weaned/100 ewes joined

Think about these statistics to appreciate what they mean. Some measures of lambing percentage ignore important facts like the number of dry/dry ewes or the wet/drys. Number 2 ignores lambs born dead, numbers 3 and 4 remove barren ewes, number 5 forgets about deaths between birth and docking.

Some farmers sell the non-pregnant ewes after scanning in midwinter and forget to add them into the calculations, so the figures they show you as a potential buyer can be misleading unless you know how to interpret them.

The true performance is probably best reflected in number 8. This is the true lambing percentage, because the number of ewes joined (i.e. presented to the ram for mating) is usually the number carried over winter, and every one of those sheep should earn its keep.

14. From Birth to Weaning

Lamb mix-ups after birth

When ewes with two or more lambs give birth in the same paddock, some lambs can be taken over by the wrong mother. Mistakes can reach about 12%, even in well-managed flocks. These errors can have important genetic implications. DNA testing is now available to sort out problems of parentage, but prevention is much simpler and cheaper.

To avoid problems of lamb mix-ups:

- Give lambing ewes plenty of space to find a quiet spot to lamb. If this is difficult, provide artificial shelter with hay bales to make quiet spots.
- When ewes have just lambed, quickly mark twins and triplets and leave them until they are bonded.
- Keep dogs well out of the way of lambing ewes unless they're completely under control. Ewes will eventually ignore a well-controlled dog that they know.
- Avoid disturbance in the lambing paddock as much as possible.
- If there are particularly popular lambing spots where mix-ups regularly occur, fence off these areas. Provide more shelter if necessary to encourage ewes to lamb over a wider area.

Watch for the burglar ewe

An occasional ewe that is close to lambing will steal a lamb from another newly lambed ewe. Their maternal instincts get out of kilter and they can be very determined – you may be tricked into thinking they have lambed when they haven't.

These ewes can be a real nuisance, as they don't produce milk until they've lambed themselves so their adopted lambs may starve.

The only solution is to shut ewes known to be burglars away from other lambing ewes until they themselves have lambed.

Deaths in the first three days

The perinatal period (the time around birth) lasts from birth to about three days of age. On hill farms up to 17% of single lambs die during this period and even more multiples, but on well-managed lifestyle farms with more intensive shepherding and more opportunity to control feeding and shelter, relatively few lambs should be lost.

There are many potential causes of perinatal death, including dystocia (difficult birth), starvation, hypothermia (exposure) and early infections such as watery mouth (see Chapter 30).

One factor contributing to dystocia is oversized single lambs. Scanning to identify single-carrying ewes, then rationing their feed in the last six weeks of pregnancy, can help to reduce this risk.

Many factors can contribute to starvation and exposure, including mis-mothering and bad weather. Again, scanning to pick the ewes carrying two or more lambs, provision of more shelter for lambing ewes and more intensive shepherding at lambing may reduce the risk.

Deaths after three days

Young lambs love to play. You see them, especially in the evenings, racing around in gangs, chasing one another across the paddock and climbing everything they can. Unfortunately in their zest for life they can end up in water troughs where they drown, or they can fall down holes (tomos) that they cannot get out of.

Before lambing, take the precaution of putting some mesh across water troughs or some large rocks in the bottom.

Cover any holes in the paddocks or fill them in.

Protection from wet and cold

Wet and cold are guaranteed killers of lambs – they cause hypothermia. In bad weather it's a good idea to put a cover on any newborn lamb that doesn't get a chance to dry.

There is a range of lamb covers on the market, from plastic to felted wool. You can even use old bread bags – cut a hole in the closed end for the head, and two smaller ones for the front legs. The bags will fall off in a couple of days. Pick them up in case stock eat them.

Helping the newborn get its first feed

After a normal delivery it can be difficult to resist the temptation to help a struggling lamb to its feet and to help it suck. But as a general rule it's best to let the ewe and her lamb work it out for themselves for at least one to two hours. If the lamb hasn't been able to find the teat in this time you could give it a hand. Tail wagging and steady sucking are good signs.

If you do help the lamb get its first feed, don't hold its head – it will struggle against this. Support the hind end so that its head is free and in about the right position to find the teat.

Colostrum

Colostrum is the first milk produced by the ewe after birth. It is an essential food for newborn lambs as it is full of antibodies needed to protect them from many common diseases.

Colostrum is thicker and yellower than normal ewe's milk, and it is so important it could well be called 'liquid gold'! Its quality decreases quite rapidly after birth, and some ewes produce better-quality colostrum than others.

A lamb should receive at least 200ml of colostrum in the first 12 hours of life – the earlier the better. After this time, the ability of the gut to absorb antibodies decreases rapidly.

Don't feed non-colostrum proteins (i.e. cow's milk) to newborn lambs because this means that subsequently they will not be able to absorb colostrum proteins properly.

Colostrum

- Helps keep the lamb warm because it is rich in energy.

- Is easy to digest and helps the lamb strengthen and grow.
- Can work miracles on weak newborn lambs.
- Helps to move the meconium (foetal faeces).

Bottle-feeding colostrum

If lambs do not get enough colostrum in their first few days they are susceptible to infections such as diarrhoea and pneumonia for months afterwards.

If the lamb is too weak to suck naturally, colostrum can be milked from its mother or from other newly lambed ewes and fed by bottle or stomach tube. Continue this for at least four days if possible. Lambs should get about 100-200ml per feed (depending on the size of the lamb) – 600ml in total in the first 24 hours of life. Larger newborn lambs need up to 1500ml daily.

If colostrum from a ewe is not available, newly calved cow colostrum can be used. You can buy good-quality colostrum substitutes, but it is important to use only the type that contains antibodies. Be wary of homemade colostrum substitutes that use egg yolks and the like. These are not suitable because they do not contain protective antibodies.

Feeding by stomach tube

Colostrum can be given to newborn lambs by stomach tube if they are too weak to suck. A rubber stomach tube specially for lambs can be bought from rural suppliers or your vet. Tubes that are too large can cause damage.

Extend the lamb's head so the mouth, throat and gullet (oesophagus) are in a straight line. Gently thread the tube through the mouth into the throat, then down into the gullet, taking care to ensure it hasn't gone into the windpipe. If the tube is correctly inserted, you will see it distend the gullet a little on the left of the windpipe as it goes down the neck into the stomach.

Warm the colostrum to body temperature before pouring it down the stomach tube. (Never warm or thaw colostrum in the microwave because the important proteins will be damaged and will congeal.)

If colostrum is not immediately available for the first feed after birth, give electrolyte solution with no protein added (see your vet).

Hand-rearing

Hand-rearing lambs is time-consuming and it can be expensive.

Orphan lambs require frequent feeds of good-quality colostrum at least four or five times a day for the first few days of life, then good-quality milk for at least six weeks. Because they are susceptible to cold, they need effective shelter, with protection from cold, wet conditions. And all this means a lot of work with no days off until they are at least six weeks old, after which time they can be gradually weaned off milk.

Milk quality

After a few days of colostrum feeding you should switch to a good-quality commercial powdered milk formulation, which must be fresh.

Cows' milk (straight from the cow or pasteurised) can be diluted and sweetened slightly to provide an effective substitute for ewes' milk.

Feeds must be small and frequent when the lamb is very young or very small, and warmed to body temperature. Giving very young or very small lambs large feeds of cold milk can cause digestive problems that result in diarrhoea.

Follow the instructions on the bags of powdered milk to be sure you are giving the correct quantity at appropriate intervals. Never make it stronger than the recommended amount: better to make it weaker.

Cleanliness is next to godliness

It's important to keep all bottles and teats scrupulously clean to prevent infections that can cause diarrhoea. Diluted bleach can be used as a disinfectant, but rinse the utensils well between feeds.

Offer hay and pasture early

Offer the lamb good-quality pasture and good-quality hay from two to three weeks of age so its digestive system develops normally. Take care though when pet lambs have access to the garden. Many have died accidentally as the result of browsing poisonous plants such as rhododendron.

Weaning

Generally, to benefit the lamb, the later you wean the better. However, the longer you leave a lamb on its mother, the more it'll compete with her for feed, and it'll also pick up its mother's worms. So it's a matter of balancing all the variables.

The amount of pasture available is a crucial factor if the lamb's potential for growth is to be realised.

Single lambs on good feed should weigh at least 25kg liveweight at weaning. Hill-country lambs will be less – about 18-20kg. This is generally at three to four months of age. Lamb growth rates between birth and weaning should be between 150g and 350g per day depending on feed.

Twins and triplets and of course quads will always be lighter. The lighter lambs are at weaning, the longer it takes to get them growing really well after weaning. They will always tend to be lighter than singles even up to the two-tooth stage.

Weaning weight targets

The 'total weight of lambs weaned' from each ewe is a good way to identify highly productive ewes. About 70% of a lamb's weaning weight can be attributed to its mother's milk; the rest to its genetic makeup.

Divide the weight of all lambs at weaning by the total weight of all ewes at mating and you have a really good measure of flock productivity.

Weaning procedure

At weaning time, take the ewes as far away from the lambs as possible so they are out of earshot. (Ewes are easier to drive away than motherless lambs not used to dogs or droving!)

Keep them both in well-fenced paddocks – away from your

bedroom and neighbours, as they'll bleat loudly for a couple of days and nights.

Ewes with very little milk will stop bleating for their lambs very quickly. They'll be glad of the respite! Put the ewes on short feed, but with plenty of water until their milk dries up. This should take two to five days.

Watch for any udders that look excessively hard, red and swollen in case mastitis develops. Ewes with mastitis may seem slightly lame on a back leg.

Discuss with your vet the testing of lambs for cobalt (vitamin B12) and selenium status.

15. Tail Docking and Castration

For just about as long as sheep have been farmed, lambs have been routinely subjected to two surgical procedures that make it easier for their owners to manage them – and make life easier for the animals too.

Castration helps curb the natural aggression of rams and prevents unwanted pregnancies, and tail docking helps keep the rear ends of sheep clean, reducing the risk of flystrike.

Castration and tail docking are potentially painful procedures but they are less traumatic if they are carried out while the lamb is young. Both procedures require skill, and owners should consult a veterinarian or experienced farmer to learn how to do the job properly.

Lambs should be castrated and docked at the same time. Use of rubber rings for both procedures when they are very young means the procedure is quick and safe and the pain appears to be soon forgotten. The risk of the lamb contracting an infection such as tetanus through the wounds that are created is reduced if the ewe has been fully vaccinated against clostridial diseases (see page 164).

Tail docking

Removing lambs' tails soon after birth has advantages for the sheep and for the farmer. The main one is that it helps prevent dirty faecal dags forming under the tail. Dags attract the blowflies that cause potentially fatal flystrike.

However, the docked tail needs to be long enough to wag! This may surprise you, but if the tail is left long enough to be raised, it lifts the supporting tissue around the anus and, like a cow's tail, directs any diarrhoea away from the body. But if the tail is amputated close to the body, any diarrhoea runs down the back end, soiling the skin and fleece. So the tail should be left long enough to cover the vulva, or the equivalent length in males (to the end of the bald bit on the underside of the tail).

Leave the tail on any male lamb that can't or shouldn't be castrated for any reason (e.g. it may have only one testicle present). This will make it obvious for culling at weaning.

Rubber rings

The most humane method of tail docking is to use rubber rings on young lambs – preferably seven to 10 days old. These are best applied with special pliers (elastrators). It has been shown that the pain of this procedure lasts a shorter time than that from tail docking with a knife.

The tail will drop off in 10-14 days and should leave a clean scar.

Check the wound from time to time to make sure it doesn't become flyblown.

Cauterisation

A gas-heated hot iron can be used for tail docking, and is especially useful for large numbers. Downward pressure on the iron cuts and cauterises the lamb's tail all in one action. However, if you press too hard and don't allow enough time, you'll cut the tail off before the dock is cauterised. This means the dock will bleed. Check the docks of the completed lambs before putting them back in their paddock.

You can spray fly repellent on the dock if blowflies are a risk, but don't dip the lamb's rear end in insecticide dip or disinfectant. The fluid gets contaminated with dung over time and does more harm than good.

Using a knife

Tails can be cut off with a sharp knife, and after a while the blood flow should stop. However, we do not advise use of this method. Lambs tend to spray each other with blood when they shake their docks after docking, and this then attracts blowflies. You can put a rubber ring on the stump to stop the bleeding, but this can be very messy.

A note for food enthusiasts: lambs' tails were once cooked as a snack for Kiwi shepherds. They threw them into a fire to burn the wool off, and the meat was tender in a few minutes. The burned wool was rubbed off and the shepherds could suck off the meat. The dogs enjoyed what was left. Maybe you'd prefer to put yours down the offal hole – but make sure you count them first to work out your docking percentage!

Castration

First of all, do you need to castrate your ram lambs? There is no need to castrate lambs that are destined to be killed before they are six months old. If you leave them entire they will grow faster than wethers, because testosterone is a natural growth stimulant. And they'll still be classed as lambs for the meat trade until their first pair of permanent incisors have appeared.

So don't castrate if you don't have to. However, make sure the ram lambs are kept separate from females from weaning at about four months – just in case!

It goes without saying that castration is very painful unless it's done skilfully. The most humane option is to have the procedure carried out by a veterinarian using sedatives, painkillers and/or anaesthetics, but for most commercial farmers the cost of this makes it impractical.

Most castration of ruminants is carried out by the farmer/owners, and it is best for the animal that the least stressful procedure be used while it is very young.

Rubber rings

For castration of lambs, the most humane method is application of a custom-made rubber ring to the neck of the scrotum with the appropriate applicator, preferably while the animal is between seven and 10 days old – and definitely before it is six weeks old.

Using rubber rings is also the easiest method, and the pain doesn't last long. The rings are the same ones used for tail docking.

Place the ring above the testicles and below the rudimentary teats. Make sure both testicles are in the scrotum before releasing the ring above them.

Using a knife (surgical)

Surgical castration (i.e. cutting the scrotum and pulling the testicles out) is another option. This is a painful operation unless painkillers are used, so ideally it should be carried out by a veterinarian using pain control.

Surgical castration is generally not a good method for lifestyle farmers to choose. It is more traumatic for the animal than using rings, and there is a higher risk of complications such as infections. The older the animal, the more potential there is for the operation to be painful and stressful.

Using a sharp sterile knife or scalpel, and keeping working conditions very clean, slit the scrotum and pull out or 'draw' each testicle in turn. They are slippery to hold, which is why, in the old days, shepherds used to 'draw' them with their teeth!

Using emasculator pliers

Castration can also be done with a Burdizzo emasculator. This implement crushes and breaks the spermatic cord without cutting the skin of the scrotum. The technique is not common, probably because of slow speed and cost of the pliers.

'Cryptorchid' procedure

Some farmers choose to 'cryptorchid' their ram lambs using the 'short scrotum' technique. This is not true castration because the testicles are not removed and testosterone production continues. But because the testicles are pushed up against the body, the animal should be infertile.

Cryptorchids benefit from extra growth from retaining the testicles that supply testosterone. Beware, though: the odd one has low fertility so they should be kept separate from ewes!

It's important to get advice from an expert, preferably a veterinarian, before attempting this procedure.

16. Slaughter

It's something most of us don't like thinking about – killing our sheep at the end of their useful lives. But although we may be inclined to keep the odd favourite as a pet, as farmers we necessarily have to send most of our sheep to the meatworks or arrange for them to be killed on the farm.

Home killing

If you want to use the meat from your sheep for your family, and if you use the services of an expert slaughterman, killing your sheep at home is the best option. If the sheep are killed at home it is easier to ensure that they are not stressed before slaughter, and this means their meat will be of maximum tenderness.

You might be able to contract the services of an appropriately MAF-licensed mobile butcher. He or she will ensure humane slaughter, good hygiene and disposal of the remains.

If you choose to slaughter your own sheep, you must get training from someone with experience in this area.

You should choose a method that does not cause the sheep undue distress and that results in instantaneous loss of consciousness leading to death. It should be a method that is reliable, safe and efficient. You must also consider the feelings of other people. Most people find slaughter of animals unpleasant, so make sure you don't have any onlookers who are not directly involved.

Shooting

A well-aimed shot into the brain with a .22 rifle or a captive bolt pistol is the best option.

There are charts showing the best site for shooting in the Code of Recommendations and Minimum Standards for the Welfare of Sheep, and the Code of Recommendations and Minimum Standards for the Welfare for Animals at the Time of Slaughter. These are available for reference from most SPCA offices and also on the MAF website (maf.govt.nz).

The site to aim for is on the forehead, slightly to one side of the intersection of imaginary lines drawn from the base of each ear to the inner corner of the eye on the opposite side. Aiming at this point should ensure destruction of enough of the brain to cause immediate insensibility and then death. Head restraint may be necessary to ensure the shot is directed accurately.

A rifle should be held about 10cm from the head to prevent back-pressure causing rupture of the barrel. A captive bolt pistol is held directly against the head.

Only licensed gun users should use firearms, although no licence is required for a captive bolt pistol.

As an added safeguard in case the shot is not immediately fatal, the throat should be cut immediately after shooting.

Sensible precautions should be taken to ensure that the sheep is securely penned and restrained throughout, and that there is no risk to operator or bystanders or other animals, e.g. as a result of a ricochet.

Throat-cutting

After shooting, the next best method of euthanasia is cutting the throat. This is the traditional method of home killing. It doesn't look nice, but it is relatively humane if it is carried out by a skilled person.

The most humane method is cutting the throat without breaking the neck. The original technique of cutting the throat and following up immediately by breaking the neck is less humane because after the throat has been cut, it takes a few seconds for the sheep to lose consciousness. Breaking the neck only causes additional trauma in the seconds before death.

The knife used should be very sharp and the blade at least 15cm (six inches) long.

The sheep should be restrained gently but firmly, standing or lying on its left side with its chin in the operator's left hand (vice versa for left-handed people), to extend its neck.

The wool over the throat just behind the angle of the jaw is parted, and one swift deep cut is made across the lower part of the neck, severing both main (carotid) arteries and both main (jugular) veins. The windpipe (trachea) and gullet (oesophagus) will also be severed.

The sheep will lose consciousness in three to eight seconds and should be held gently but firmly until then. A huge amount of blood will be lost in a few seconds.

The signs of death (as opposed to unconsciousness) are lack of pulse and breathing, pupils of the eyes widely distended, and lack of blinking reflex when the surface of the eye is touched.

Cutting the throat is *not* a humane method of euthanasia in any species other than sheep or goats as the blood supply to the brain of other species is different and the time to lose consciousness will be longer.

Preparation for the meatworks

For many of the veterinary medicines and agricultural chemicals used on livestock there is a 'withholding time'. This is the length of time that must elapse before meat or wool from treated animals is sold. Make sure these limits are observed before selling wool from any treated sheep or sending it for slaughter.

All sheep presented for slaughter should be accompanied by an animal status declaration (ASD) form listing the animal remedies they have been given in the last 60 days. Other information, e.g. about Johne's disease vaccination and concentrate feeding, also needs to be supplied. You can get these forms from meatworks but if you sell through an agent he or she may liaise with the meatworks for you.

Contact several meat buyers or stock agents well before the time to find out where you will get the best deal. You might be best to sell directly to the meatworks, or to meat buyers at the saleyards.

In deciding which works to use you must balance various factors. Price is a big consideration, but you probably also want to minimise stress on your sheep. This means sending your sheep to works that are as close to home as possible.

Make sure the sheep are clean and try to choose a works that does not routinely swim-wash sheep that arrive in clean condition. Swim-washing means they are forced through a 20-metre race of cold water, and this is a stressful procedure carried out on some or all sheep in every export meatworks in New Zealand.

If you keep the sheep in a bare yard for 12 hours before transport they will be 'emptied out', meaning they will be less likely to defaecate on one another in the trucks and yards. You should also trim off (crutch) any dirty wool and dags around the rear and belly of the sheep before they leave the farm.

It is illegal for the transport operator to load any animal that is not fit for the journey. It must be able to stand and bear weight on all four limbs, and it must be strong enough to withstand the rigours of the journey without undue discomfort or pain.

No sheep with a broken leg should be transported. A written veterinary certificate must be obtained when any doubt exists about the fitness of the sheep to travel.

Do your homework, and choose the best option in terms of both financial return and your sheep's welfare.

Prime lambs

Prime lambs are an important money-earner so they should be in top condition when they go for slaughter. The best time to send your lambs is when they reach the optimum liveweight and condition. Consult with your buyer. The prices are paid according to the lambs' carcase weight, and in lambs this is about 38% to 50% of liveweight.

You can assess the trace element status of lambs still on the farm by arranging for tests on liver samples taken from the first group to go to the works. See your vet.

Cull ewes

If you slaughter your cull sheep and intend to sell the meat, you must use slaughter premises appropriately licensed by MAF.

Many of your culls will have worked hard for you over the years, so you owe it to them to make their last journey as stress free as possible. Try to avoid selling your ewes to an agent who will put them through saleyards before they go for slaughter. This process can take three or four days and it is very stressful, particularly for old, thin sheep.

Like prime lambs, culls should be presented for slaughter in good body condition for maximum return. Since they are usually in poorer condition than prime lambs to start with, the culls will need preferential treatment in the days and weeks leading up to slaughter.

Swim-washing is likely to be particularly stressful for thinner weaker sheep.

Any sheep that is not fit for transport can be humanely killed on the farm. You may get some money for the carcase if you get a MAF-approved petfood operator to slaughter the sheep for you on the farm. The carcase can then be taken to MAF-licensed petfood premises for processing.

17. Wool Production

Wool prices have been low for many years, and an interest in woolless sheep is on the rise – again! But wool is still a wonderful product, and although you may not make much money from it, you can maximise your returns by treating it properly.

Sheep should be shorn at least once a year (unless they are one of the woolless breeds). Sheep with more than a year's wool growth can suffer heat stress. They can become cast if their fleece gets waterlogged, and their fleece will become dirty and daggy, making them more prone to flystike.

When to shear

There are many different shearing regimes:

- The whole flock once a year, usually in spring or early summer
- Ewes in midwinter (called pre-lamb shearing)
- Ewes in mid-pregnancy and then again later in the year
- The whole flock twice a year (e.g. in winter and summer)
- Three times in two years – when the wool is an acceptable length

Preparing sheep for shearing

- To prepare sheep for shearing, have them crutched and dagged if they are dirty at the back end (see page 116).
- In the weeks leading up to shearing, don't use any raddle crayons that are not clearly labelled as 'scourable'. Even then, use them sparingly, especially aerosols.

- Keep sheep off pasture in a big yard for at least 10 hours before shearing to allow them to empty out. This helps prevent their dunging on one another's wool, and they will be more comfortable during shearing if their bellies are not full.
- Never shear wet sheep, as it will cause health problems for shearers (boils), and damp wool will heat and go mouldy in store.
- Before shearing a pet lamb, check that its collar has been removed!
- Don't treat sheep for external parasites in the weeks before shearing – treat them after shearing. The withholding periods for strong, mid-micron and fine wools are 60, 90 and 120 days respectively. Chemical contamination of wool is a serious marketing problem.
- If your fences are in poor condition, check fleeces also for bits of wire before shearing. They break combs and injure shearers.

Remember that for up to six weeks after shearing, sheep require from 20 to 70% more feed than before (depending on the weather) and they benefit from good shelter.

Ensuring fleece quality

If you're going to sell wool it must be well prepared and there is considerable skill involved in this. Aim to present the best-quality fleece wool (from the main body of the sheep) as the main line. This wool is what the market wants, and it should not be contaminated with lower-quality wool from other parts of the body.

During shearing:

- Keep all synthetic fibres (e.g. baler twine and string) away from sheep and their wool. It will contaminate the wool and it wrecks wool-processing machinery.
- Make sure shearers and shed hands don't leave their singlets and towels near wool presses and bales. If they get into the wool they'll wreck processing machinery and cause massive contamination, costing thousands of dollars' damage when the wool is processed!

- Have obvious bins in the shearing shed for all rubbish and make sure they're used.

When the shorn fleece is on the floor or table, remove the poorer-quality wool from it before putting the good stuff into the wool bale. This means removing:

- All dags (if the sheep haven't already been dagged and crutched)
- Belly wool
- Wool from around the crutch and from the lower legs
- All raddled (paint-marked) wool
- All wool stained by faeces (including the 'pen stain' that results when sheep are kept too close together in the yards or shed)
- Top-knot and cheek wool
- Any fribs (sweat locks) around the crutch and under the front legs

Length of wool is important in marketing. Generally it should be 75-100mm (3 to 4 inches) long. Prices for very long wool and very short wool are lower because it's not suitable for modern high-speed processing machinery.

A major disruption in the sheep's feed supply or other stress (disease, lambing, bad weather) can cause a 'break' in the wool. The wool grown at the time of the stress is much finer than normal so the fibres break easily during processing. You can detect this if you pull on either end of a wool staple (a 'lock' of fleece) and it gives or comes apart. Wool that has a break may still sell but will be discounted.

Winter shearing

There are benefits from winter shearing. The ewes' metabolic rate increases, so they eat more. As a result they are less susceptible to metabolic disease and their lambs tend to be bigger and more robust. The ewes produce more milk and they seek shelter in bad weather so their lambs benefit.

But there are risks from winter shearing. Don't shear thin sheep (with a body condition score of less than 1.5). Winter-shorn sheep must be given up to 70% more good-quality feed than before, and they must have access to shelter in bad weather. Covered yards with sheds should be available for emergency shelter.

If you shear in winter, use combs that leave a bit more wool on the sheep as protection against bad weather. These are called winter combs, cover combs, lifter combs or snow combs. In the South Island some high-country sheep are still shorn with blades to leave more wool on.

When pre-lamb shearing ewes, consider leaving their belly wool on. It means they will be more likely to lie down to rest, and then their bulk can provide useful shelter for their lambs.

Selling the wool

You can sell your wool through a wool merchant, who will buy it from you on the spot, or you can take it to a merchant who will sell it on your behalf (although this may take some weeks). Check your Yellow Pages under wool buyers.

A wool merchant usually pays you less for small lots because of the extra handling charges needed to bulk them into a saleable line of usually four bales.

18. Preventative Health Care

Crutching

Crutching means trimming off the wool around the crutch of the sheep. The crutch is the area immediately around and below the tail, down between the hind legs and halfway along the underside of the body.

Urine, faeces and vaginal discharges can soil the crutch. Keeping wool around this area short helps keep the sheep clean, prevents flystrike, and helps newborn lambs find the teats.

If you have only a few sheep you can use hand-held wool shears or blades. For bigger numbers it's easier to get contractors to use shearing machines.

To do a full crutch, start with the sheep sitting up comfortably against you on its rear end (tail), and remove the wool from inside the right leg. Then trim around the crutch and along the left leg. This is more easily done with a machine than blades.

In a ewe the full crutch includes the wool around the udder – and certainly all the wool between the udder and the vulva. Be very careful not to slice bits off the udder, teats or vulva.

In a ram, a full crutch includes trimming any long wool off the scrotum and around the pizzle (the penis sheath). Again take care not to cut the skin, the pizzle or your fingers. Pull the skin taut with your left hand while shearing. Do not trim hair from the end of the pizzle during crutching. This hair helps to drain drips away, preventing pizzle rot.

To complete a full crutch, the sheep is turned to sit on its side, and the wool is trimmed down each side of the rear of the back leg, around the anus and vulva, and a bit over the tail head. Press your left hand into the lower loin of the sheep just in front of its stifle (knee) to make its back legs go straight. This reduces the chances of cutting skin or tendons (hamstring).

A 'ring crutch' is a much smaller version of a full crutch. It involves trimming wool from over the tail head and around the anus and vulva, but none from between the back legs.

Make sure your shearing blades are clean and sharp. Also all shearing combs and cutters should be clean, sharp and properly adjusted for operator safety and to avoid skin cuts. If you have an old shearing plant and handpiece, have it checked by someone with experience. A faulty handpiece can cause serious injury if it comes apart or jams while in operation.

Always keep a first aid kit in the shearing area.

Dagging

Dagging means removing all the dirty wool around the rear end and belly of a sheep. Dags are clumps or aggregations of soft or hard faecal material tightly bound to wool, usually under and around the tail, anus and hind legs. Sometimes mud can form dags too, especially around the belly.

Blowflies can attack skin under both kinds of dags but prefer a bit of moisture. If sheep get daggy, remove the dags when they are still soft, and before the blowflies find them.

Dags can be removed by careful use of hand shears, but it can be difficult to get the blades under the dags without cutting the sheep. A set of dagging shears with very short blades makes the job easier. Learn to sharpen your hand shears or blades properly, and keep them spotlessly clean at all times.

If there are more than a few sheep to clean up, it's easiest to use a shearing machine handpiece. Use old thin combs as it's easier to get under the dags, especially if they're dry.

Don't expect shearers to dag your sheep before shearing – dags spoil their shearing combs and stain the shearing board, which then stains the wool clip.

Drenching

Internal parasites (worms) are probably the most common sheep ailment. Worms commonly cause ill-thrift and big numbers can be fatal. Regular drenching to control worm burdens is usually essential, and this is dealt with in detail Chapter 20.

Dipping

Lice can be a problem on some farms, and lice problems are treated using insecticide.

The term 'dipping' is sometimes used by farmers to refer to any process of getting insecticide onto the sheep's wool because at one time the commonest method was physically dipping the sheep in a bath containing insecticide. Nowadays 'dipping' generally includes applying pour-on treatments and spraying.

More details are given below in Chapter 30 under 'External parasites'.

Foot trimming

A sheep has a 'cloven' hoof, which means it walks on two digits (compare this with a horse, which has one, or a dog, which has four).

Between the two digits, sometimes called 'claws' or 'toes', is the 'interdigital space', which can easily get filled with mud and small stones. This is also a great place for infections such as foot scald to develop.

The walking surface of each toe should be flat, but in soft or wet conditions where there is no natural wear, the horn ('toenail') from the sides can grow long. Sometimes it grows so long it curls up, making it hard for the animal to walk. Or the horn on the two digits gets so long they cross over.

The excess horn should be trimmed as necessary, usually about once a year. The key point when trimming feet is to cut only dead horn (like our own toenails), not deep into sensitive tissue containing blood. If you draw blood when you trim feet, you will probably cause painful infection to develop.

Foot-trimming equipment needs to be kept sharp and very clean, with regular disinfection.

Vaccination

Vaccination against common diseases is a good insurance policy against losses.

Vaccination of all sheep on the farm with a '5 in 1' vaccine against the main clostridial diseases (tetanus, pulpy kidney, blackleg, black disease and malignant oedema) is recommended routine procedure. It is readily available, relatively cheap and very effective.

It may be wise to vaccinate against other diseases too, such as scabby mouth, Johne's disease, salmonellosis or the infectious abortion diseases (campylobacteriosis, salmonellosis and toxoplasmosis) depending on the incidence of these diseases in your area. Ask your local veterinarian for advice.

More information about vaccination is given in the appropriate disease sections (Chapters 24, 26, 27 and 30).

Trace element supplementation

On many farms the soil is deficient in cobalt (converted to the vital vitamin B12 in the body) or selenium, and some areas are deficient in copper or iodine.

Sheep on these farms will need to have their diet supplemented regularly to ensure that trace element deficiencies will not result in production losses and disease (see Chapter 25).

Your veterinarian will advise on testing your stock to find out what supplements they need and the best way to provide them.

Keeping your property disease free

This chronicle of sheep diseases might tend to put you off keeping sheep – but don't despair! Remember the simple principle: Prevention is better (and cheaper) than cure.

Keeping disease out of your flock is basically a question of good husbandry (or wifery!). It's important to try to keep your farm disease free as far as possible, because once you have disease organisms on your farm, they may be there for many years:

- Be very careful from whom you buy stock – choose established and reputable sheep farmers.

- The demands of 'traceability' driven by meat companies mean that it will get easier to get evidence of stock quality from professional farmers.
- Have your boundary well fenced against neighbours' stock.
- Consider double-fencing the boundary (with trees for shade and shelter in between), so stock are not in contact with the neighbour's stock. It might seem an extravagance, but it would benefit your stock.
- Don't allow stock trucks onto your property. Unload on the roadside or in an area reserved for the purpose.
- Establish a quarantine area where all new stock are kept for a period of a few days after arrival and after treatment with anthelmintic to reduce the risk of them bringing disease onto your farm.

19. Sheep Health Calendar

Prevention is so much easier and cheaper than cure. Here's a look at the sort of preventative programme that will see your sheep thrive and grow with a minimum of ill-health.

Ewes before lambing

- The ewe needs a 5-in-1 vaccination one month before lambing. The immunity produced in the lamb will last for about 16 weeks.
- This is to protect the lambs (provide immunity) against tetanus, pulpy kidney, blackleg, black disease and malignant oedema.
- Ewes need to be crutched before lambing to remove wool from around the tail and udder.

Lambs at birth

- Lamb in clean paddocks.
- Disturb ewes as little as possible in the first couple of days after giving birth.
- Put iodine on all fresh navels, but watch the smell doesn't cause mismothering.
- Make sure all lambs have had dams' colostrum.
- Mark twins and triplets with raddle to check correct mothering.
- Make sure multiples are mothered up correctly before dark each night.

- In bad weather make sure newly lambed ewes and lambs have shelter.
- In rough weather, put multiples inside if possible for the night or in the best shelter.
- About 3-5 days after birth, make sure dried faeces are not blocking lamb's tail areas.
- Where possible, get rid of afterbirths lying round the paddock.
- Bury or dispose of all dead animals as soon as possible.

Lambs 3–6 weeks old

- Dock all lambs' tails with rubber rings or cauterising iron.
- Castrate male lambs with rubber rings.
- Use of Insect Growth Regulator (IGR) spray-on if docking late lambs near flystrike season.
- Vaccinate for scabby mouth if advised by your vet.
- At docking vaccinate all lambs with 5-in-1 if dam has not been vaccinated before lambing.
- At docking, consider drenching ewes if advised by vet.

Lambs 12–16 weeks old

- Wean lambs. They will grow better on clean pasture on their own.
- Drench at weaning for round worms and tape worms. Check with vet for suitable drench.
- Give ewe lambs that are to be retained in the flock a booster vaccination at weaning with 5-in-1. This immunity should last till their pre-lamb vaccination.
- Provide good feed on clean worm-free pastures.
- Provide good water and shade in hot weather.
- Shear lambs if part of shearing policy. Shearing will lower flystrike risk.

Weaned lambs

- Avoid the stresses of unnecessary yarding and handling – especially in high temperatures and dusty conditions.
- Provide good green feed to achieve high growth.
- Provide good water and shade in hot weather.

- Drench for internal parasites. This is a critical period, especially when autumn rains arrive. Check with vet for appropriate product to be used and to arrange faecal egg counts.
- Ask your vet about testing your lambs to find out whether they need cobalt (vitamin B12) or Selenium supplementation, and arrange supplementation if necessary.
- Dip lambs 2-3 weeks after shearing. Pour-ons are most practical for small flocks.
- Watch for blowfly, even when lambs are not dirty.
- Except in the South Island, start facial eczema prevention (zinc). Zinc bolus is the most practical method.

Hoggets (14 months)

- Dag all dirty hoggets before shearing.
- Shear hoggets (last year's lambs) at about 14 months old, before current crop of lambs are weaned.
- They will show a growth spurt after shearing if you provide good green feed.
- Provide good water and shade in hot weather.
- Dip (spray or pour-on) 2-3 weeks after shearing.
- Never dip sheep for at least 60 days before shearing.
- Watch for blowfly strike, even when sheep are not dirty.
- Ask your vet about testing for trace elements and arrange supplemention if necessary.

Hoggets (18 months)

- If hoggets are well grown (min 40kg), you could consider joining them with the ram.
- Join them 2-3 weeks before main flock rams go out.
- Continue facial eczema prevention until May.
- Feed hoggets well all through autumn and into winter to maintain growth.
- Pregnant hoggets are priority stock – give them the best feed on the farm to keep them growing.

In-lamb hoggets (22–24 months)

- Vaccinate pregnant hoggets 4 weeks before lambing (5-in-1 vaccine).

- Provide hoggets with extra attention at lambing.
- Provide good quality feed to stimulate good lactation.

Two tooths

- Give two tooths special care as they will usually be lambing for the first time.
- If they lambed as hoggets, they will be easier to manage.
- Lambing hoggets need to be fed extra well as they are still maturing.
- Give two tooths a booster 5-in-1 vaccination after mating (if not vaccinated as lambs post weaning) and then a booster vaccination before lambing.
- Ask your vet about testing for trace elements and arrange supplementation if necessary.

Older sheep – annual jobs

Shearing

- Shearing regimes vary, the most common being once a year, and twice a year for strong wools.
- Dag all sheep before shearing – shearers will not shear dirty sheep, or if they will, you will be charged extra.
- Neither will shearers shear damp sheep. Make sure they are perfectly dry.
- Provide good feed and shelter for all sheep after shearing.

Dipping

- Do not dip during the 60 days before shearing.
- Dip (spray or pour-on) about 2-3 weeks after shearing, and keep a close watch for blowfly attack at any time of year.

Vaccinating

- Vaccinate ewes with 5-in-1 vaccine a month before lambing.

Drenching

- Only drench mature sheep for internal parasites if a faecal egg count justifies it.
- Discuss with your vet the option of drenching ewes when docking their lambs.

- Discuss with your vet the option of drenching ewes before lambing.

Feet care

- Treat all sheep for footrot as soon as possible to prevent spread on the farm.
- Isolate severe cases. They are hard to cure. Consult your vet.
- Foot scald often occurs in lambs and ewes when grass is long, lush and wet.

Udders

- Check ewes at lambing for mastitis. Affected ewes will be dull, in pain and often limping in a back leg. Udder will be swollen, red and hot to touch. The milk will be clotted and smelly.
- Check ewes frequently at weaning until their milk supply dries up.
- Check teats are not cut or injured at shearing.
- Check the udders and teats are correct before putting the ram out.

Teeth

- Make sure all ewes going to the ram have sound mouths (ie. a full set of good incisor teeth). Sheep do have not top front teeth and bite against their top gum.
- If you keep ewes with missing incisors they will need longer pasture.
- Use only rams that have sound teeth and incisors that meet the top pad at its front edge (not in front or too far behind).

Reproduction

- Don't breed from ewes that have had bearing problems.
- Get rams checked by your vet a month before mating each year.
- With your vet, check selenium status and provide supplementation if necessary.

20. Drenching

Worm control

Gastrointestinal worms are without doubt one of the biggest threats to the health and welfare of grazing animals in New Zealand. These internal parasites live in the stomach and intestines and lay eggs, which are passed out in faeces and hatch into larvae on the pasture.

If the larvae of worms from one particular species of animal are eaten by animals of the same species, they grow into adults and begin the life-cycle again. The worms can produce thousands of eggs in every gram of faeces passed. It's a good lifestyle as far as the worms are concerned, because they are very successful and there are lots of them! But it's not so good for the animals or their farmers.

Unless farmers take steps to prevent it, worm burdens in livestock can build up to unmanageable levels. Sheep with heavy worm burdens develop diarrhoea, become unthrifty, pot-bellied, anaemic, lose their appetite and may die from having hundreds of thousands of worms in their digestive tract. By the time worm numbers in the intestine have become large enough to cause diarrhoea, body condition and growth rate will have been adversely affected.

There are conditions and diseases other than worms that can cause these signs, so if there is no response to drenching, consult a veterinarian. For example, lush spring grass can cause diarrhoea, and so can some bacterial infections in the intestine.

Freezing and drying out tend to kill larvae, so pasture contamination can drop during cold winters and dry summers, but generally worm burdens build up to their highest levels in animals in autumn.

Sheep are generally most susceptible to worms while they are young, but fortunately as they age they tend to build up a natural resistance.

Types of drench

As a general rule, on most farms regular effective drenching of sheep at strategic times throughout their lives with anthelmintic (medication to get rid of internal worms) is absolutely vital for their good health.

There are three types of drench (or drench families) available:

- White drenches (benzimidazoles). These act quickly to kill worms in the sheep's stomach and intestine, but do not kill any worms from a day or so after drenching. This means lambs need to be drenched at four-week intervals from weaning until about May.
- Clear drenches (levamisole). The clear drenches kill worms quickly, with no lasting activity. They appear to be less effective than the white drenches or third-generation drenches; for instance they do not kill larvae that have lodged in the stomach wall (ostertagiasis). Clear drenches should be used only at weaning and in early summer. In autumn, when the worm burden on pasture tends to be at a height, a white drench or third-generation drench should be used.
- Third-generations (ivermectin and related types). Third-generation drenches or endectocides (e.g. ivermectin, abamectin, doramectin, moxidectin and spiromectin) are more expensive than the white or clear drenches but they continue to kill worm larvae in the animal for about two to three weeks after drenching. This means the interval between doses can be extended to six or eight weeks, which makes the cost per day comparable with other drench families. Some of the third-generation anthelmintics also kill some types of lice. These are called endectocides.

The type of drench you use is one of the factors that determines your drenching programme through summer and autumn.

Anthelmintics now come in various formulations. Most are given by mouth through a drench gun but there are also injectable anthelmintics, as well as long-acting capsules that sit in the rumen producing anthelmintic for around three months. (Considerable care and skill are required to make sure the capsules are administered correctly without damaging the animal's throat.) Note that there are no pour-on anthelmintics as there are for cattle and deer.

The drench families can be rotated on a farm each year to avoid a buildup of drench resistance in internal parasites, which is a huge problem on New Zealand farms.

Keep in touch with your vet for the latest recommendations regarding the best drench to use.

Vets and other advisers are now tending to recommend programmes of minimal or 'strategic' drenching, using faecal egg counts to work out how often to drench. This has the advantage of ensuring that drenching is cost-effective and it reduces the chances of developing drench resistance.

No place for guesswork

It is very important to follow the manufacturer's instructions carefully so that you get the drenching technique and the dose right. Make sure the appropriate full dose of drench is given. If you are giving a standard dose to a group of lambs of roughly the same size, choose the correct dose for the heaviest.

Under-dosing encourages the survival of worms that are resistant to the type of drench used.

Immediately after drenching, the sheep should if possible be moved to pasture that has not been grazed by that species for some time, e.g. after hay or silage has been taken off and pasture has regrown.

Drench resistance

Drench (anthelmintic) resistance is a huge and growing problem on livestock farms, particularly with sheep and goats.

If you have drench resistance on your farm it means that some of the worms on your pasture and in your sheep or goats have developed resistance to a particular type of anthelmintic, so drenching with anthelmintics in that drench family will not be effective in getting rid of these worms.

Once you have drench resistance in the worms on your farm, it is very difficult or impossible to get rid of it.

Testing for drench resistance/drench effectiveness

Tests on faecal (dung) samples 10 days after drenching will tell if the drench was effective or not. The tests will show how many eggs there are per gram of faecal sample. This measure is called the faecal egg count or FEC.

If the drench was effective, there should be no worm eggs present.

If there are worm eggs present, the worms may be resistant to the drench used, or the drenching technique may have been faulty (the drench gun may not have been working properly), or insufficient drench may have been given for the size of the animal.

Your veterinarian or adviser will explain how to collect the dung samples and will arrange for laboratory testing, which is inexpensive and quick.

If drenching was not completely effective, your adviser will help resolve the problem.

If drenching is not effective

If drenching is not effective in reducing faecal egg counts to zero, then either you are not drenching properly or you have drench resistance.

Check your drench technique, i.e. make sure you are drenching properly (see below).

If you have drench resistance you will need to drench again using an effective drench from another drench family (see below).

Drenching properly

It is important when drenching to give each sheep a full dose for its body weight, because underdosing encourages survival of resistant worms.

If the sheep in the group you are about to drench are very similar in size, you can give them all the same dose, but it must be the correct dose for the heaviest of them. You will need to weigh a few of the biggest to find which is the heaviest, work out the correct dose for this heaviest and give this dose to every sheep in the group.

If the sheep in the group are very varied in size, weigh each animal before dosing to make sure it gets its full dose.

If the sheep are small enough for you to lift easily, you can weigh them using bathroom scales. Carry the sheep in your arms and get someone to weigh you with the animal, then subtract your weight.

You must make sure your drench gun is delivering an accurate dose by testing it before use. Draw up a calibrated dose, 'drench' an empty syringe barrel and check that the dose it delivers is the calibrated dose.

Don't bring resistant worms onto your farm

Drench resistance can be brought onto farms in introduced stock.

Brought-on sheep and goats should be drenched with an effective drench immediately on arrival.

Ideally, they should be held in a quarantine area for a few days and released onto the farm only when their faecal egg count is zero.

Combine drenching and good pasture management

The best worm control programme for your farm will incorporate rotational grazing, mixed species grazing and strategic dosing with reliable anthelmintics.

You will need a veterinarian or an experienced and competent farmer to help you devise this programme. A well-designed programme will definitely pay dividends in terms of improved stock health and welfare, and productivity.

'Natural' or organic methods

Most farms, and this includes most small farms, don't have the space or flexibility to be able to rely on natural parasite control methods, although natural methods are an important component of every worm control programme.

Natural methods include grazing paddocks in rotation, allowing time for some or most of the worm eggs and larvae to die off on the pasture between periods of grazing.

Grazing pasture with different species in turn can help reduce pasture contamination, because most types of worm are species-specific. (Note, however, that goats and sheep share the same internal parasites.) For example, the larvae of horse worms will die when eaten by a sheep, so following horses with sheep in a paddock, or following sheep with ponies, can reduce the numbers of worm eggs and larvae on pasture. Low stocking rates also tend to mean lower pasture contamination.

Since mature sheep are more resistant to worms than lambs, graze pasture first by lambs, then put older sheep on it.

Alternatives to anthelmintics, such as garlic and cider vinegar, can sometimes help suppress the egg-laying of worms, and may even remove a few worms, but they do not get rid of them and cannot be relied on as a means of worm control.

Quarantine drenching

There is a risk that any sheep arriving on or returning to your farm could introduce worms that are resistant to some anthelmintics.

You need to ensure that they are effectively drenched to eliminate their worm burden before they are released onto your main paddocks. Put aside an area for this purpose – this will be your 'quarantine paddock' or 'quarantine yard'. The paddock can be grazed by cattle or horses as well as drenched newly introduced sheep but never by home sheep or goats.

After drenching sheep with a third-generation drench, hold them in this paddock for a few days or until faecal egg counts are zero (see below), providing water and if necessary supplementary feed.

PART THREE
Diseases and Other Health Problems

21. Flystrike

In flystrike, blowflies lay eggs on the skin. Maggots hatch from the eggs and eat into the skin and flesh, causing sores. Affected sheep suffer irritation, pain, loss of fluids, infections and eventually they die.

Flystrike is a sickening sight, even for the most experienced farmer. It is a cruel disease that costs the sheep industry almost $40 million a year and it occurs throughout New Zealand.

Of all domestic animals, sheep are most often affected, because wool – particularly dirty wool – attracts blowflies. One type of blowfly, the Australian green blowfly, can strike even relatively clean areas on sheep, and it has a longer season than other species.

Blowflies favour warm humid conditions, and dirty sheep and flystrike are most common in late summer and autumn.

Signs

You'll need to be very observant to spot the early signs of eggs on the wool or skin or small sores containing maggots. The sores are usually under patches of wet or dirty wool at the sheep's back end, in shearing cuts or on feet with footrot. The Australian green blowfly can also strike relatively clean skin along the back or around the poll and ears.

Sometimes the behaviour of the sheep is the first sign that it has been struck. It might seem restless, seek shade, twitch its tail, swing around to try to nibble affected areas or stamp its feet.

Treatment

When there are sores on the skin, the maggots should be removed with meths (a horrible job). The sore should then be treated with flystrike powder containing diazinon, available from your vet or rural supplier. In an emergency, small flystrike wounds can be treated with flyspray.

It's important to treat the sores when they are still small. Extensive and deep sores are very difficult or impossible to treat. In severe cases, euthanasia may be the only humane option.

Affected sheep should be examined daily to make sure they don't become struck again. Insect repellant on the surrounding wool will help to keep the blowflies away.

Prevention

Dirty, wet or injured skin attracts blowflies, so it's important to remove the attraction by keeping sheep clean:

- Removing dags is important, and if the wool is long, shear the sheep to prevent the wool at the back end getting wet and dirty. Sheep should never be carrying more than a year's growth of wool.
- Good worm control helps prevent the diarrhoea that soils the wool at the back end.
- Blowflies favour sheltered areas so move sheep to open more exposed paddocks during high-risk periods.
- Blowflies breed in dead carcases as well as on live flesh, so it is very important that all dead animals on the farm are buried or incinerated. This includes wild animal and bird carcases too.
- Flytraps can help attract blowflies away from stock. If enough flytraps are used early in the season, they help prevent flystrike. Flytraps can also be used to monitor the risk by indicating how many blowflies are around. These can either be bought or made at home, and many vets and agricultural advisers can advise on how to make effective traps.
- Pour-on or spray-on treatments are also effective in prevention e.g., after docking. There are many control treatments on the market and it is important to select one

suitable for your situation. Read the label carefully and follow the manufacturer's instructions to the letter. Each treatment has different specifications. Make a note of how long they will last. For maximum effect, treatments should be applied soon after shearing.

- Following treatment there is a withholding time before wool can be sold. Note this carefully.

22. Facial Eczema

Facial eczema (FE) is a type of photo-sensitisation or exaggerated sunburn that can affect sheep, cattle and goats. It can also affect deer, especially fallow deer.

Affected animals are uncomfortable, may be in pain, and usually look miserable.

The skin damage is secondary to liver damage, and both together cause ill-thrift, reduced fertility, drop in milk production and sometimes death. There is always liver damage with FE.

In all but the southern two-thirds of the South Island, facial eczema is one of the biggest health and welfare problems affecting sheep and cattle. The fungus that causes FE favours moist warm summer and autumn conditions, so FE is most prevalent in parts of the North Island, and it can appear from January to May.

Signs

The first signs are usually reddening and swelling of skin exposed to the sun – the skin around the eyes, ears, lips and nose. Then sheep become restless, shaking and rubbing their head and ears, and seek out shade. The skin reddens, swells, weeps, then it sloughs off and the site can become infected.

Animals with severe liver damage can appear jaundiced, with a yellowish tinge to the whites of their eyes.

Cause

Facial eczema is caused by a toxin (sporidesmin) produced by a fungus (*Pithomyces chartarum*). The fungus grows in the dead litter at the base of the pasture in warm, moist conditions.

When the spores are eaten by ruminants, the toxin damages the liver. The liver then cannot get rid of phylloerythrin, a chlorophyll-breakdown product, and this circulates in the blood. Phylloerythrin releases energy when exposed to sunlight to cause skin damage identical to severe sunburn.

Monitoring the risk

If facial eczema is a problem in your district it is vital to keep up to date with current spore counts. Some veterinary practices display up-to-date spore count charts in their reception areas.

Spore numbers in pasture samples can be measured using a simple microscope technique that is relatively easy to learn. Some farmers set up their own monitoring groups. You could join one.

Spore counting is an invaluable way to help assess the risk in your paddocks and take preventative steps in good time.

Techniques have been developed recently to monitor FE spore density in faecal samples, and this is a more direct measure of how many spores have been eaten. Ask your veterinarian about this.

Prevention

Use spore monitoring to indicate the degree of risk. Don't wait until you see the first clinical cases before you act. By the time 5% of a herd or flock have obvious skin damage, up to 50% of the group will be liver-damaged and at real risk of developing skin signs too.

The most effective way to prevent FE is to give zinc oxide by mouth. Depending on the circumstances, drenches can be given daily, several times a week or once a week. It's important to check the dose rates and mixing recipe carefully with a vet or farm adviser at the beginning of the season – as early as December. Zinc supplements should be started in January.

Alternatively, sheep can be given intra-ruminal long-acting boluses. These last for 28 to 42 days and are less labour intensive. Special care needs to be taken in administering these boluses as they can seriously damage the animal's throat if you use undue force. Put stock out onto grass as soon as the bolus has been given, just as you'd take a drink after swallowing a big pill to help it go down.

It is important not to risk overdosing by doubling up on prevention methods.

Stocks of zinc can run out when demand is high, so put your order in early. Even when stocks of zinc oxide are low, do *not* drench with zinc sulphate. This can be irritant and toxic and can kill dosed animals. To prevent adverse reactions, farmers should use only licensed products. Note that there are currently no licensed injectable zinc products on the market.

Keep sheep off pasture that is likely to have high spore counts. They could be moved to bare land or yards and fed winter reserves such as hay, at least until the spore counts drop. This should be a practical option on many small farms. However, there are risks in concentrating stock on bare land or short pasture for too long as the combination of high stocking density and stress can predispose to diseases such as salmonellosis.

Pasture can be sprayed with fungicide. This is relatively expensive but it may be effective in high-risk areas. The manufacturer's instructions should be followed carefully, and to test the effectiveness of spraying it is important to monitor spore counts on sprayed pasture.

Long-term solutions

FE resistance is strongly inherited, so buy in FE-resistant stock, and use FE-resistant rams that have a certificate to prove their status. There are ram breeders who have been breeding for FE resistance for 20 years or more, and some of them have a very high level of protection built into their flocks. (This means their sheep show fewer ill-effects after grazing pastures with high spore counts than do less tolerant sheep.) Buy your rams from these breeders.

Consult your vet to discuss the pros and cons of blood-testing sheep to help select replacement ewe hoggets and rams.

This can be relatively expensive but it may be cost-effective on some farms.

The blood GGT test measures the amount of liver damage done during the season, and selecting sheep with low levels helps promote FE resistance in the flock.

Most newly imported breeds have less resistance to FE than New Zealand breeds. The Finn is an exception.

Treatment

Farmers have a moral and legal obligation not to allow affected animals to suffer unnecessarily. They must be treated for facial eczema. There is no cure for the liver damage caused by sporidesmin. Though the liver has a great capacity to recover from damage, some lasting damage is likely to remain.

At the first signs the affected sheep should be taken off pasture and offered shade. A dose of zinc oxide should be given by mouth. Jaundiced stock should be offered a diet of hay and water for a few days before gradually introducing high-energy nutritious feed to help the liver recover.

Opaque protective cream can be applied to damaged skin to hasten healing and to screen the skin from sunlight. Your vet can give antibiotic injections if severe skin infections occur, and also vitamin B12 injections to boost appetite.

Jaundiced sheep should not be slaughtered or sent for slaughter for human consumption.

The importance of shade

FE is a form of sunburn, and getting affected animals out of the sun provides relief and helps prevent further damage. Affected animals often seek the shade of trees or hedges, but this may not be sufficient. During high risk periods or when severe signs appear suddenly, for example swelling of the head and ears, stock can be kept in darkened, well-ventilated housing during the day to let the worst of the reaction subside.

They can be offered hay and water and let out at night to graze. Where there is appropriate housing this is a simple, cheap and effective way to prevent or help treat FE.

Consult the experts

Preventing FE is much better than trying to deal with FE-affected stock, as many farmers learn to their cost each autumn. If you are uncertain what to do, talk to an expert right away. Many veterinary practices and farm consultants offer their clients a practical and topical advisory service.

23. Ryegrass Staggers

Ryegrass staggers is a brain disease of sheep, cattle, horses and ponies, deer and alpacas. It occurs commonly throughout the North Island and as far south as North Otago. It is most common in summer and autumn, especially when stock are forced to eat down to the base of the pasture.

Cause

The disease is caused by the ingestion of a toxin called lolitrem B, produced by a fungus in perennial ryegrass. The fungus is an endophyte, meaning it grows within the living plant. The highest concentrations of toxin are in the leaf sheath at the base of the pasture and in the seedheads.

The toxin has a specific damaging effect on cells in the cerebellum of the brain that coordinate movement.

Signs

The signs are most obvious when affected sheep are disturbed and forced to move. They become anxious about being approached. In mild cases there is slight trembling of the head and of the skin of the neck, shoulder and flank.

More severe cases show head nodding and jerky movements, swaying while standing and staggering during movement.

In the most severe cases animals have a stiff-legged gait, short prancing steps, and may collapse with rigid spasms that last for up to several minutes.

Hazards for affected sheep

The disease itself is not fatal, but there is a real risk of injury or death as a result of accidents because:

- Affected sheep lose weight as they don't graze as much.
- They may not be able to drink sufficient water.
- They can become caught up in obstacles such as electric fences.
- They can fall into holes and ditches, and over bluffs.
- They can drown in creeks, dams, drains and swamps.

What can you do?

Remove sheep from hazardous pasture. There is no cure, but you can make sure the sheep come to no harm while the effects wear off.

If safe pasture is not available, put the stock into yards and feed them hay (or silage/baleage) and plenty of clean water.

Handle affected sheep quietly and don't disturb them unnecessarily.

The best long-term solution in areas where the disease is a problem is to replace the affected ryegrass pasture with species of ryegrass that are endophyte free or contain only safe endophytes such as ARI.

24. Pregnancy Problems

Metabolic diseases in ewes

In late pregnancy and early lactation, ewes are under great metabolic stress. Their foetuses grow fast in late pregnancy, and after giving birth they have to produce a lot of milk.

If their feeding is interrupted, for example by bad weather or by yarding, they can easily be tipped into fatal metabolic imbalance. The result may be one of the following metabolic diseases:

- Hypomagnesaemia (grass staggers - not to be confused with ryegrass staggers)
- Acetonaemia (also called sleepy sickness or pregnancy toxaemia or twin lamb disease)
- Hypocalcaemia (milk fever)

In all these diseases the first signs are usually a change of behaviour. It may be a dullness, progressing to the stage where the ewes go down and are unable to rise, or it may be agitation with trembling and nervousness leading to convulsions.

It's wise to discuss treatment options with a veterinarian before cases develop, and vital to call a vet at the first sign of trouble. Emergency treatment is essential if the animal is to survive.

Other predisposing factors include poor body condition, a check in feed supply, cold wet windy weather and the stress of transportation. To prevent metabolic diseases, try to keep the feed supply steady or increasing, provide sheltered paddocks in bad weather and minimise the time stock spend in yards.

Hypomagnesaemia

Hypomagnesaemia (low magnesium in the blood) causes occasional deaths in lactating ewes on lush pasture, although it can occur in dry sheep as well. It is often caused by excessive application of potassium and ammonium fertilisers.

The early signs can be subtle, with an increase in nervousness that can be easily overlooked or misinterpreted. The nervousness can lead to mis-mothering lambs.

It is sometimes accompanied by dramatic signs such as body tremor, walking with stiff legs, collapse with paddling and the head held back. Usually ewes are just found dead on their sides, with scuff marks in the ground where they have been kicking.

Blood tests will tell when blood magnesium concentrations are getting low, and urgent treatment is then vital. Treatment involves injection of magnesium solutions, preferably by a veterinarian.

Prevention includes providing shelter from bad weather, and dusting pasture or hay with magnesium oxide powder or calcined magnesite in spring if the risk is significant (e.g. there is a history of hypomagnesaemia on the farm). Dusted pasture may not be very palatable, and if the powder is applied too liberally the stock may go hungry to avoid it. This can precipitate acetonaemia. Wear a mask when applying the dust.

Acetonaemia

This is the most common metabolic disease in sheep. In ewes it is called sleepy sickness or pregnancy toxaemia or twin lamb disease. It occurs in the weeks before lambing.

Ewes carrying two or more lambs are especially at risk, and so are very thin or very fat ewes.

The main causes are underfeeding or a sudden check in food intake in the last six weeks of pregnancy. It can also be brought on by inadequate shelter in bad weather, when food intake is reduced but feed requirement is increased.

Unlike the other metabolic diseases, the onset of sleepy sickness is not usually sudden. The signs include slowness, lethargy, not eating, staggering or aimless wandering, twitching of the face and ears, blindness leading to recumbency, usually with the head up.

Coma and death can follow in two to seven days.

Ketones are excreted in the urine and in the breath. These have a characteristic sweet smell (like nail varnish remover!) which about 50% of people can detect around affected animals. The blood concentrations of ketones rise before signs develop, so blood tests can be used by your vet to predict problems.

The key to prevention is good feeding. Ewes should have an increasing level of feed leading up to birth, and concentrate feed may be necessary.

Acetonaemia in animals that have been underfed for some time is generally very difficult to treat because of liver damage. There are various energy supplements - some with electrolytes - that can be given by mouth, and they can be helpful if given very early in the course of the disease.

Hypocalcaemia

Milk fever is caused by low calcium concentrations in the blood. It occurs in older ewes in the weeks around lambing (from six weeks before to eight weeks after lambing).

The disease is often brought on within 24 hours of a sudden stress. This might be yarding, transportation, forced exercise, very bad weather or insufficient feed.

Sometimes a sudden change of feed, such as a move to lush pasture, will trigger outbreaks.

The signs are restlessness, trembling, staggering, depression and recumbency. It is characteristic to find ewes down on their chests (rather than on their sides) with their hind legs extended out behind them, and head down and extended forward.

There may be a discharge from the nose, they may become bloated and they usually abort dead lambs. Affected ewes may just be found dead.

Treatment of milk fever is by injection of calcium solutions at the first sign of problems.

Abortions

It can be devastating for the sheep farmer to go out on a winter's morning to find aborted lambs in the paddocks. Such a waste! And it makes things worse to learn then that most types of abortion (or 'slips') can be prevented by vaccination. There are effective vaccines available against toxoplasmosis, campylobacter and salmonellosis, all of which can cause abortion in sheep. Vaccination should be carried out well before mating or early in pregnancy.

Even if abortions begin there are steps to take that can lessen the impact.

Abortions can occur at any stage of pregnancy, although usually only mid- to late-term aborted foetuses are big enough to be noticed.

Sometimes the aborting ewe is unwell. If this is the case, or if it looks like more than an occasional spontaneous abortion, consult a veterinarian as soon as possible. There are various possible causes. The most common are toxoplasmosis (caused by a tiny parasite), campylobacter infection and other types of bacterial infection such as salmonellosis.

Toxoplasmosis is spread from the faeces of cats, which can excrete the eggs for a short time, usually when they are kittens. (Older cats don't usually pose a risk to sheep, so don't heed old wives' tales about getting rid of farm cats to prevent toxoplasmosis.) The ewe does not usually show any signs of ill health.

A sensible precaution against toxoplasmosis is not to feed hay to younger, more susceptible pregnant ewes if it comes from barns where there may have been kittens. This hay can be fed to non-pregnant hoggets and can help them develop immunity for future years.

Some types of salmonellosis cause abortion in ewes, and affected ewes will become very dull and may have severe diarrhoea. They will require veterinary treatment. *Salmonella* Brandenburg has caused major abortion outbreaks in the South Island in the last few years.

Poor-quality silage and baleage can contain bacteria (Listeria) that cause severe diarrhoea and abortion in ewes.

If abortions start, spread ewes out to reduce the risk of infection spreading. Pick up aborted foetuses and membranes and bury them.

Be good to your ewes, since stress seems to precipitate more abortions.

Bearings

A bearing is a mass of flesh bulging from the vulva of a heavily pregnant ewe. It may consist of the inside-out vagina, or it may be bigger, containing the bladder and/or cervix and/or womb. It is a very ugly sight and it can cause the ewe considerable distress.

Several factors contribute to bearings, including high pressure in the abdomen as a result of a womb full of lambs, a rumen full of frothy herbage, a lot of fat in the abdomen and a full bladder.

Prevention

The risk of bearings can be reduced by ensuring a steady supply but not an overabundance of quality pasture.

Keep at-risk ewes off hilly land as lambing approaches, and perhaps encourage them to take gentle daily exercise.

There may be a genetic component to susceptibility to bearings, so it is best to cull ewes that have survived bearings, and not to keep their offspring.

Treatment

It is important to check all ewes in late pregnancy at least daily for the first sign of bearings. It is easier to treat a bearing that is small or very recent. It is very difficult to treat successfully when it has dried out and become damaged.

Where the bearing recurs or when it is very large or infected, the most humane option may well be euthanasia.

It is important to use only humane methods to replace and retain bearings. Treatment involves very gently cleaning and replacing the prolapsed vagina. It is important to hold the prolapse up to allow the bladder to empty before trying to replace the vagina. Facing the ewe downhill or elevating her hind end can help. Lubricant is useful too.

If there is any doubt about what to do or how to do it, you must call a veterinarian. The ewe will need antibiotic injections anyway, to prevent serious womb infections.

Once the bearing has been replaced, it must somehow be retained or it will prolapse again. Commercial plastic bearing retainers can be inserted into the vagina and tied to the wool to keep them in place. But the gentlest method of retaining prolapses involves tying wool over the vulva. Or string can be tied to the base of a 'finger' of wool on one side of the vulva and stretched tightly across to the base of a finger of wool of the other side. Two or three such ties may be sufficient.

Another simple and often effective technique involves using a length of baling twine. This is placed across the back in front of the pelvic bones (pin bones), brought down the sides of the ewe, passed inside each hind leg, crossed over the vulva and tied back onto itself where it goes over the ewe's back in front of the pin bones. The string should be sufficiently tight to prevent the prolapse from recurring.

There are other methods of retaining bearings, some of which involve stitching or using metal clips to keep the vulva closed. This type of procedure is only acceptable if it is carried out skillfully and hygienically. If in doubt, call in your vet.

Remove retainers before lambing

Although ewes can sometimes push lambs out past retainers, the retainers should be removed just before lambing begins, and stitches or clips must be removed before lambing.

Shearers don't like finding retainers or twine in the fleece. Make sure all of it has been removed.

25. Trace Element (Mineral) Deficiencies

If they are to perform well, all sheep need small amounts of certain vital nutrients, called trace elements or minerals. In many parts of New Zealand, soils and pastures are deficient in the minerals cobalt, magnesium and/or selenium, and in a few areas copper or iodine are lacking.

These nutrients - albeit in only tiny amounts - are vital for the good health of sheep. Any sheep dependent on pasture deficient in any of these minerals may develop deficiency-related ill-thrift or disease.

Growing lambs and pregnant sheep therefore often need supplementation. Consult your veterinarian to find out whether your sheep need mineral supplementation and how to give it.

Vets can arrange blood and liver tests and soil and pasture tests if necessary. You can also arrange for tests to be carried out on liver samples collected from the first lot of lambs sent to the works.

Cobalt

Cobalt deficiency can cause ill-thrift in all ruminants, but particularly in lambs. Loss of appetite leads to poor growth, weight loss, wasting and eventually death. There may be a watery discharge from the eyes, and scabby ears. Cobalt deficiency is also associated with an increase in the occurrence of brain disease and liver disease in older sheep, and an increase in newborn lamb deaths.

Cobalt deficiency in sheep can be combated by injecting vitamin B12. Long-acting vitamin B12 injections are now available.

Cobalt can also be added to fertiliser and topdressed onto pasture.

Selenium

About a third of New Zealand's soils are deficient in selenium. Selenium deficiency can cause ill-thrift, especially young growing lambs. It can also cause white muscle disease in lambs, and has been associated with infertility in ewes.

Selenium can be added to fertiliser and topdressed onto pasture, or added to anthelmintic drenches. Long-acting selenium injections are now available for sheep. In areas of mild deficiency a salt lick containing selenium may help, but the only way to ensure every animal gets an adequate dose in areas of significant selenium deficiency is through drenches, injections or topdressing.

Copper

Copper deficiency is more common in peaty soils and leached sandy soils. Livestock species vary in their susceptibility to copper deficiency, with deer and cattle generally more susceptible than sheep. Some other minerals in the diet, such as molybdenum, sulphur and iron, inhibit the absorption of copper.

Deficiencies result in hind leg weakness, fragile bones, poor growth rates, infertility, reduced fleece weight and loss of wool.

Copper deficiency is unlikely to be a problem on most lifestyle farms, but if necessary, it can be added to the food as a supplement or injected. Long-acting copper oxide wire particles can be given by mouth, or copper can be added to fertiliser and topdressed onto pasture.

Magnesium

Magnesium deficiency is caused by low pasture concentrations of magnesium, often caused by excessive use of potassium and ammonium fertilisers. High concentrations of potassium and nitrogen in spring pasture also contribute to the problem of magnesium deficiency.

The high demands of pregnancy and lactation can result in insufficient magnesium for pregnant and lactating ewes. Ewes that are not well fed are especially at risk. The signs are increasing agitation before convulsions and death. Bad weather or yarding or any other interruption of grazing in heavily pregnant animals predisposes to the disease (see Chapter 24).

Magnesium can be added to the diet – sprinkled on hay or dusted onto pasture – but it has to be given daily during the high-risk period.

Iodine

In a few inland areas and alluvial plains, soils are deficient in the trace element iodine. Iodine deficiency can also be induced by feeding brassicas and clovers, which contain chemicals (goitrogens) that reduce thyroid hormones.

One of the main signs of iodine deficiency is goitre – a swelling of the thyroid glands in the neck just below the throat. Occasionally, goitre occurs in lambs.

Other signs of deficiency include poor survival rates in newborn lambs, reduced wool production and reduced fertility in ewes. Diagnosis is generally based on the occurrence of these clinical signs.

Ewes and lambs can be dosed with potassium iodide at intervals of three to six months. A long-acting iodine injection is also available.

In lower-risk areas, providing iodised salt licks may be sufficient.

If your area tends to be iodine deficient, discuss supplementation options with your veterinarian.

Don't overdose

While selenium and copper are vital nutrients, only tiny amounts are required, and larger doses can be poisonous. Even two or three times the recommended dose of either of these minerals can be toxic.

For example, don't give selenium to sheep in drenches if it

has already been topdressed onto the paddocks.

Some breeds of sheep, like Texels, are more susceptible to copper poisoning than other breeds, and some breeds, e.g. Finnish Landrace, are more susceptible to copper deficiency than other breeds.

It is important to seek veterinary advice before providing livestock with selenium or copper supplements.

26. Scours

Yes, sheep get runny poos too! Diarrhoea (scours) in sheep means that their normal faeces, which should resemble brown peas, become softer; they might get runny or watery and even blood-stained.

There are many causes of diarrhoea – from irritants in the food to infections.

The signs are soiling and maybe dagginess of the wool around and below the anus.

Eventually, because of loss of nutrients and fluid into the intestine, persistent severe diarrhoea causes dehydration and the sheep may die. The soiled wool can also attract blowflies, leading to flystrike.

Worms

Worms are the most common cause of scours in sheep. Worms in the intestine and sometimes in the stomach can reach huge numbers and cause severe diarrhoea and even death (see Chapters 18 and 20). This is the most common cause of dirty backsides on sheep, and it's particularly likely to cause problems in young sheep before they develop much age-related immunity.

Coccidiosis

There is another type of internal parasite that can cause diarrhoea – Coccidia. These are tiny one-celled organisms that they live in the lining of the intestine. They too can reach huge numbers, causing severe dark brown/black diarrhoea in lambs up to a year old, called coccidiosis.

Sometimes the disease sweeps through the mob without causing too many problems, but if it's severe, veterinary treatment is needed or affected sheep may die.

Salmonellosis

There is a group of Salmonella bacteria that can cause very severe diarrhoea in sheep, usually mature sheep. *Salmonella* Typhimurium and *S.* Brandenburg are among them.

Veterinary treatment is imperative, because the disease is often fatal and can spread to other types of animal, including humans.

The signs are severe watery and blood-stained diarrhoea, and affected sheep are very dull. There is a vaccine available.

Yersiniosis

One type of bacterium causes yersiniosis, usually in lambs in their first winter. Yersiniosis can be a relatively minor transient disease causing only soft faeces for a week or so, but occasionally it is more severe and requires veterinary treatment with antibiotics.

Viruses

Viruses such as rotavirus, particularly in housed or hand-reared lambs, can cause diarrhoea in young lambs up to a few weeks of age. Seek veterinary advice.

'Spring scours'

When there is no other obvious cause, 'spring scours' in lambs is sometimes thought to be due to irritants in pasture. Some fungal toxins growing in the dead vegetable matter at the base of the pasture in autumn can cause ill-thrift and diarrhoea.

With the help of your vet, it's important to rule out other possible causes, such as internal parasitism, and if the problem persists, move sheep to different pasture.

27. Feet Problems

Footrot

Sometimes the horny wall of the toe separates from the sensitive internal structures and the space gets filled by dirt. A bacterium that lives in the soil can become established and quickly cause footrot.

If your sheep are lame and on close inspection you see that the horn has separated from the wall of the foot and it smells rotten, the sheep probably has footrot.

If the horn overgrows, it makes footrot more likely.

To treat footrot, first cut back any overgrown horn, being careful not to draw blood. Use a clean sharp pair of clippers.

Apply an approved aerosol spray – check with your vet – then put the sheep through a footbath containing 10% zinc sulphate. Stand the sheep in the footbath for at least 5 minutes.

Less effective is a 4% formalin solution. Formalin is nasty stuff to deal with – take care when you handle it. Stand the sheep in the formalin solution for a few minutes, then let them stand on concrete for half an hour so that the formalin dries on the feet. Note that too many formalin foot baths (more than twice a week) can damage the feet.

Eradication of footrot can take a long time and you need to tackle it with a vengeance, but it will be worth it if your sheep are very susceptible. Seek veterinary advice on the best products for your situation, and how to run an eradication programme.

In the end you may need to cull sheep with persistent problems, and pay special attention to the feet of any rams you use. Susceptibility to some foot diseases is partly genetic.

There is a footrot vaccine available – see your vet to find out if it's appropriate for your sheep.

Foot scald

Foot scald shows as red inflamed or blanched white areas between the claws. It's a bacterial infection common on lush pastures and is likely to cause lameness, sometimes so severe that the sheep is reluctant to stand. It can be just as painful as footrot and can develop very quickly.

Smell the foot – foot scald doesn't stink the way footrot does.

A good soak in a footbath containing 10% zinc sulphate is generally an effective treatment.

Foot abscesses

Foot abscesses are nasty swellings on the feet. They are also caused by bacterial infection, they contain pus and they make sheep very lame. Rams with foot abscesses may be reluctant to mount ewes. Remove these rams from the flock.

Seek urgent veterinary advice, as any affected sheep will need antibiotic injections, and they take time to heal.

Cull affected animals, as the problem may recur.

28. Teeth Problems

It is difficult to examine the premolars and molars (the back teeth) in sheep, but the incisors are relatively easy to examine. The most common problems of incisors are excessive wear and periodontal disease.

Excessive tooth wear

With excessive wear, the incisors become very short and may be worn down to the level of the gums. This happens when soil or sand on pasture gradually files away at the teeth over time, so it is usually only a problem with older sheep.

Sheep with very worn incisors can fare well – as long as they don't have to eat very short or stringy pasture. They can be fed long soft pasture or hay or silage or concentrate feed.

Periodontal disease

Periodontitis or periodontal disease can cause the incisors to grow long and loose, eventually falling out. When there are missing incisors the sheep has a 'broken mouth'.

If there are only a few very loose teeth they are best removed, either with your fingers or with pliers. Then the sheep will be 'gummy', but it might fare well enough if it is offered long soft pasture or hay or silage or concentrate feed.

Cheek teeth

Sometimes the gum and supporting structures around the roots of the cheek teeth (molars and pre-molars) become infected. In severe cases the bone around the roots of the teeth becomes swollen and sore. The sheep is reluctant to chew its cud and it gets thin. Culling is the only option as there is no cure.

Overshot and undershot jaw

Sometimes lambs are born with a lower jaw that is too short, i.e. it meets the top pad far behind its front edge. This is called undershot jaw or 'parrot mouth'.

Sometimes the lower jaw is too long and protrudes. This is called overshot jaw.

In either case, the lamb might be able to suckle but it will probably not be able to graze properly. If you put it on long pasture, it might be able to eat well enough to grow to slaughter weight.

Don't buy breeding sheep with these jaw faults as they can be hereditary.

29. Poisons

Accidental poisoning of sheep occurs regularly. Usually it could easily have been prevented with a little know-how.

There are too many different signs of poisoning to list them all here, but the most common signs include diarrhoea, regurgitation of rumen contents, unusual excitement or dullness, body tremors, pain (teeth-grinding, reluctance to move, arched back) and convulsions.

If you suspect poisoning, consult your veterinarian without delay. In the meantime, the following precautions will minimise the risk:

- Don't throw garden prunings into the paddock. Many plants are poisonous to sheep (see below) and many are more tasty when wilted.
- Even cut grass can cause severe indigestion if sheep gorge on it.
- Fence off rubbish dumps and check native scrub for tutu and ngaio before allowing sheep access to it.

Poisonous garden plants and weeds

Rhododendron
Yew
Laburnum
Oak (acorns)
Delphinium
Blue lupin (a fungal toxin in lupins can cause lupinosis)

Iceland poppy
Cestrum
Oleander
St John's wort
Tutu
Ngaio
Ragwort (sheep are more resistant to ragwort poisoning than cattle or horses but their liver can be damaged if they eat enough of it)
Foxglove
Goat's rue

Poisons in pasture

- Some of the most serious diseases of sheep are caused by fungal toxins in pasture (see ryegrass staggers, facial eczema).
- Zealalenone is a fungal toxin produced by a fungus that grows in pasture and it can cause infertility in ewes.
- Some clovers can cause dermatitis.
- Phalaris grasses can cause signs of brain disease (phalaris staggers).
- Some rapidly growing pastures, especially after nitrogen application, can contain toxic quantities of nitrate.
- Immature rape can cause skin lesions.
- Nitrate concentrations in rapidly growing turnip tops, rape and some new pastures can be toxic, causing death.

Algae

- Algal bloom in stagnant water (blue-green algae) can cause kidney damage, with incoordination and death.

Chemical poisons

Sheep can be overdosed with zinc (see Chapter 22), or with copper or selenium (see Chapter 25).

Superphosphate poisoning can occur when sheep are put onto topdressed pasture before the fertiliser has been washed into the soil, especially when pasture is short.

Basic slag has caused poisoning when grazed before it had

been washed into the soil.

1080 poisoning has occurred in sheep when poisoned bait was accidentally dropped onto their pasture or when they got access to poisoned land.

Overdosing with organophosphate insecticide or anthelmintic can cause toxicity.

30. Other Common Ailments

Clostridial diseases

These include pulpy kidney (which causes sudden death, usually in lambs), tetanus and the gangrene diseases (malignant oedema, blackleg and black disease), which are almost always fatal.

Fortunately vaccines against these five diseases are very effective, and routine vaccination of sheep with a '5 in 1' vaccine is the norm on New Zealand farms.

Vaccination of ewes against the clostridial diseases such as pulpy kidney and tetanus is also good insurance against losses in lambs, because lambs are passively protected by antibodies in their mothers' colostrum for up to three months after birth.

It is best to vaccinate ewes twice while they are hoggets, then give them a booster vaccination each year about a month before lambing. This ensures good levels of antibodies in their milk to protect their lambs.

Lambs from vaccinated ewes do not need to be vaccinated until they are two to three months old as the ewes' colostrum contains antibodies. If ewes have not been vaccinated, their lambs should be vaccinated as soon as possible, i.e. at docking and again four to six weeks later.

The most common vaccination procedure is to actively immunise all lambs from about 10 to 12 weeks of age with a sensitiser dose, followed by a booster dose four to six weeks later.

Ewe hoggets entering the breeding flock and all older ewes in the flock are then given a booster dose before lambing.

Scabby mouth

Scabby mouth (also called orf, contagious ecthyma, contagious pustular dermatitis) is a viral disease that causes crusty sores mainly around the lips of lambs up to six months old.

Thistles and gorse seem to predispose animals to the infection, probably because they cause scratches that then become infected.

Scabs can also form around the muzzle, ears and lower legs, and around the teats of ewes. In most cases the sores heal in a few weeks without treatment. However, if the sores become extensive and severe enough on the lamb's mouth it may be unable to suck. If sores on the ewe's teats are severe enough she may be reluctant to let lambs suck, and she may develop mastitis.

There is a vaccine available that can be given to lambs at tailing. See your vet for details. There is no treatment, and the sores will heal eventually. Antiseptic creams can help prevent secondary infections and can aid healing.

Sores can develop on goat skin too, and on human skin, so always wash your hands after dealing with affected sheep.

Johne's disease

Johne's disease causes progressive weight loss leading to emaciation and death over a period of weeks or months. It most commonly occurs in sheep from two to five years old.

It is caused by a bacterium that lives in the soil and can persist for years. There is no cure, and if you are unlucky enough to have Johne's on your farm, you will need the help of your vet to develop a programme of culling and strategic vaccination to get rid of it.

Johne's disease is more common in sheep in the South Island than the North.

Stresses such as too little feed, cold weather in winter and pregnancy often seem to precipitate the disease in late winter/ spring.

Pneumonia

Acute pneumonia can cause sudden breathing difficulties and death, sometimes with coughing. In the more chronic forms of pneumonia and pleurisy, especially in hoggets, there may be few obvious signs apart from reduced thrift and reduced returns from the meatworks.

Various types of micro-organism work together in the lungs to cause pneumonia.

Sudden severe outbreaks can follow mustering in dry dusty yards in hot conditions, or shearing, particularly if newly shorn sheep get cold and hungry.

Affected sheep can be treated with antibiotics but this is not always successful.

Preventative measures include damping down dusty yards or mustering early in the morning if sheep have to be yarded in summer, and providing good shelter and extra feed after shearing.

In individual sheep, clumsy administration of drench into the lungs instead of down the gullet can cause pneumonia that is usually fatal.

External parasites

Parasites that live on the skin are called ectoparasites, and the most common of these are lice.

Lice are species-specific. Like other farm animal species, sheep can be infested with their own type of louse, and it affects sheep only.

Lice lead to damage of the skin and fleece because they cause an itch and the sheep rubs against solid objects. If you see woolly sheep with strands of wool pulled out they may be lousy.

If you have keen eyesight you may be able to find lice on the animal. Part the fleece over the itchy areas. You may see the lice scuttling for cover.

Infestations tend to be highest during winter, particularly when sheep are underfed or have big internal worm burdens.

Other parasites that live on the skin can cause irritation and hair or wool loss, and can contribute to production losses. These include the mange mites and ticks.

Sheep keds are another type of external parasite but they are rare in New Zealand.

Mange is not common. The mites that cause it are microscopic and live in or on the top layers of skin causing hair loss, sores and irritation.

Ticks are about 2 mm diameter and can swell up to 10 mm diameter when engorged with blood. They tend to be most common in the northern parts of the North Island but have occasionally occurred in the South Island. They don't usually cause serious problems in their host unless they are present in big numbers, when they can cause anaemia.

Treatment and prevention of ectoparasites usually means keeping stock fit and sturdy so that they are more resistant to infestation. For specific prevention and treatment, insecticide can be applied in the form of powders, pour-on liquids or sprays. Many of these come in a pour-on or spray-on form that is convenient for lifestyle farmers to use.

Plunge dips and spray dips are alternative options, usually provided by a contractor.

Generally it's best to use insecticides soon after shearing when the wool is short.

There are so many products available it is best to get advice from an expert regarding the most suitable for your stock and your situation.

Remember that there is usually a withholding period before the wool or meat can be sold, so bear this in mind when planning treatments. Read the label on the product.

Watery Mouth

This is a painful disease of very young lambs (up to 72 hours old). The signs are excessive salvation and distended abdomen, and affected lambs usually die. It seems to occur when conditions at the lambing site are dirty, and the lamb swallows bacteria before it has had a good feed of colostrum. Oral antibiotic given within a few hours of birth can prevent it but treatment once signs have developed is not usually successful.

Index